modern abstract algebra

modern abstract algebra

volume two modules, linear endomorphisms and algebras

YUTZE CHOW
University of Wisconsin
Milwaukee, Wisconsin

GORDON AND BREACH SCIENCE PUBLISHERS
New York London Paris

Gordon and Breach Science Publishers Inc.
One Park Avenue
New York, N.Y. 10016

Editorial Office for the United Kingdom

Gordon and Breach Science Publishers Ltd.
41-42 William IV Street
London W.C.2

Editorial Office for France

Gordon & Breach
7-9 rue Emile Dubois
Paris 75014

Library of Congress catalog card number 76-26736.
ISBN 0 677 05370 3.
 Printed in Great Britain.

to Wanlin

Preface

"Modern algebra" is now a standard part of a mathematician's discipline. The present book is written on the level of a senior or first-year graduate course. It is my hope that the book will also be useful as a reference for some mathematicians and theoretical physicists. For this reason, many discussions and comments are placed in the "Remarks" to reduce unnecessary mingling of mathematical expressions with sporadic verbal statements so that the book may be conveniently used as a reference even after a student finishes it as a textbook. Furthermore, at the end of each chapter there is a "Problem Section" where the problems are either completely solved, half solved or hints are given. This, I hope, will reduce a student's fear of problems and will enhance the success of the use of the book for self-study. It is also important to point out that many of the problems are parts of the proofs of some theorems in the text; by giving the complete or partial solutions or even hints it will enable a student of lesser background to go on without missing some crucial information.

The style of the book is based upon my intention of building up mathematical maturity gradually from objects of simpler algebraic structures. For instance, the "monoid" structure serves the purpose of introducing the notion of "free" objects while group theory provides the ground for homology.

I have made special efforts to use precise notations and terminologies in the initial introduction of any definition or concept. Such a precision is usually relaxed as the discussion developes, to ease the burden of symbols and language. Equivalence of viewpoints, conceptual generalities and the interplays of different algebraic structures are often emphasized.

The necessity of omitting certain important topics often presents a real challenge to an author. But, a textbook cannot be an encyclopedia and I can only hope that the other aspects of the book may compensate for this. In spite of the homological flavors in the general approach to many topics, we have dispensed with the use of categorical and functorial language to avoid an excessive build-up of jargons. In my personal opinion, the notion of handling things "en masse" by categories and functors can always be learned later on.

I am most grateful to Professor E. C. G. Sudarshan for his kind arrangement for me to spend a most pleasant and stimulating time at Syracuse University, where the writing of this book was initiated. Without his sustaining encouragement and interest this manuscript could never have been completed. My special thanks are also due to Professors Karl Krill and Richmond McQuistan for their kind interest.

In the preparation of the book's manuscript I have benefited from many friends and colleagues of mine.

Particularly, I want to thank Professors Kuo-Tsai Chen and Bruno Harris for their valuable advice toward improvement and their encouragement. My colleagues Professors Richard Gantos and Dale Snider have kindly read the earlier version of §.1.1 through §.2.5 and made many useful suggestions. I am particularly grateful to Mrs. Thelma Lubkin for reading the first four chapters of the final manuscript; the joint effort of Mrs. and Professor Elihu Lubkin leads to a number of useful improvements.

My wife, Wanlin, typed most of the final manuscript. My daughter, Eleanor, checked the final typed version and compiled the table of contents.

Yutze Chow

Milwaukee, Wisconsin

Acknowledgement of Copyrights

The author expresses his sincere appreciation to Annals of Mathematics for their permission to use substantial quotations of the following papers:

S. Eilenberg and S. MacLane, Annals of Mathematics, 48 (1947) 51 and 326.

Oystein Ore, Annals of Mathematics, 32 (1931) 463.

Symbols and Notations

The following is only a partial listing of the symbols and notations used in the book. It often happens that a symbol or a notation may have a different meaning, in the text, than the one which is given below. In such cases explicit statements regarding the symbols or notations will be made to avoid ambiguity.

SYMBOLS

$\emptyset$	Empty set
${}_{c}S$	Complement of the set S
$x \in X$	x is a member of X ($\notin$, non-membership)
$\forall$	Every
X	Cartesian product
$\cap$	Intersection
$\cup$	Union
$\supset$	Inclusion ($\supsetneq$, proper inclusion; $\not\supset$, non-inclusion)
$\exists$	Existence ($\nexists$, non-existence; $\exists_{1!}$, unique existence)
$\longrightarrow$	Mapping
$\longmapsto$	Element-wise mapping
$\Longrightarrow$	Implication ($\not\Longrightarrow$, non-implication)
$\Longleftrightarrow$	Two-way implication
$\longleftrightarrow$	Isomorphism or bijection
Ω	Composition rule
$\otimes$	Tensor product
$\oplus, \dot{\Sigma}$	Direct sum

	Equivalence relation
$A \equiv B$	$A=B$ by definition or choice
$\hat{1}_X$	Identity mapping on X ($\hat{0}_X$, zero mapping)
1_X	Unit element of X (0_X, zero element)
$\overset{-1}{f}$	Preimage of a mapping f
f^{-1}	Inverse of a bijection f
$A \parallel B$	A is a factor of B ($\nparallel$, not a factor)
$A \prec B$	A is a normal subgroup of B
$A \precneqq B$	A is a normal proper subgroup of B
$A \subset\!\!\!\rightarrow B$	A is an ideal of B
$A = B$, mod X	A mod X = B mod X
$A = ((X))$	A is generated by X
$K[t]$	Polynomial ring on t with coefficients in K
$\langle , \rangle$	Bilinear form

<u>Notation</u>

$\mathbf{C}_g$	Conjugate mapping w.r.t. g
$\mathbb{C}$	Set of all complex numbers
$\mathbb{C}^n$	$\mathbb{C} \times \cdots \times \mathbb{C}$ (Cartesian product of n copies of $\mathbb{C}$)
I	Set of all integers
I_+	Set of all positive integers
I_-	Set of all negative integers
$I_{(+)}$	Set of all non-negative integers
$\underset{\sim}{K}$	Represents either $\mathbb{C}$ or $\mathbb{R}$.
$\mathbb{Q}$	Set of all rational numbers
$\mathbb{R}$	Set of all real numbers

When used in homology, the following notation has the meaning:

B^n	Set of all n-coboundaries (B_n, n-boundaries)
C^n	Set of all n-cochains (C_n, n-chains)
H^n	n-th cohomology group (H_n, n-th homology group)
Z^n	Set of all n-cocycles (Z_n, n-cycles)
δ^n	n-coboundary operator (δ_n, n-boundary operator)

Abbreviations

AA	Associative algebra
adj.	Adjoint mapping
auto	Automorphism
Aut.	Set of automorphisms
Bij(X,Y)	Set of all bijections from X to Y
card.S	Cardinal number of S
cen.G	Center of G
char.K	Characteristic of K
deg.	Degree
Der.V	Set of all derivation mappings on V
det.	Determinant
dim.	Dimension
endo	Endomorphism
$End_K V$	Set of all K-endomorphisms on V
epi	Epimorphism
Epi(X,Y)	Set of all epimorphisms from X to Y
hom	Homomorphism

Hom(X,Y)	Set of all homomorphisms from X to Y
im.f	Image of f
Inj(X,Y)	Set of all injections from X to Y
Inn.$\mathcal{L}$	Set of all inner derivations on $\mathcal{L}$
iso	Isomorphism
Iso(X,Y)	Set of all isomorphisms from X to Y
ker.f	Kernel of f
mono	Monomorphism
Mon(X,Y)	Set of all monomorphisms from X to Y
Out.$\mathcal{L}$	Set of all outer derivations on $\mathcal{L}$
Tr.	Trace of a matrix

We call special attention to the following abbreviations and symbols:

iff = if and only if.

LHS = left hand side.

RHS = right hand side.

▌ denotes the end of a remark or the end of a step in a proof.

▌▌ denotes the end of a complete proof.

The word "eigen", used as a modifier in the text, is separated from the noun that follows it.

Contents of Volume Two

Contents of Volume One

CHAPTER IV

On Modules

Summary

This chapter starts with definitions of modules over a group or a ring. The topics covered here include such notions as quotient modules, chain conditions, composition series, simplicity, semi-simplicity, homomorphisms, free modules, tensor products, exact sequences, projective and injective modules. Modules over a ring are the main concern of this chapter. The special case of modules over a field (i.e. vector spaces) is only briefly mentioned here; their properties, especially in regard to their linear endomorphisms, are the central topics reserved for a later chapter.

§. 4.1. Modules over a group or a ring

In contrast to a group or a ring, the structure of a "module" involves two sets (that they may be identical is not excluded). The concept of a module, however, should not really be new to the reader; it is essentially an abelian group with a left (or right) operator set which is either a group or a ring (especially a field or a division ring) or a set with some other algebraic structures (cf. §.2.17, especially expressions [296] and [306]).

(def) Left module over a group

Let G be a group (with Ω_*). Then an abelian group M (written additively) is called a left G-module (i.e. a left module over G) with a module composition $\Omega_\square$ if

$$\Omega_\square : G \times M \longrightarrow M \qquad [1]$$

is a mapping satisfying the following conditions for every x, x' ε M and g, g' ε G:

i) $g \square (x + x') = (g \square x) + (g \square x')$ (right distributivity) [2]

ii) $(g * g') \square x = g \square (g' \square x)$. (associativity) [3]

(def) A left G-module, M, is called a unitary left G-module if

$1_G \square x = x$, for every x ε M, (left neutrality) [4]

where 1_G is the unit element of G.

The "right" axioms are introduced in a similar manner. M is called a right G-module (i.e. a right module over G) with a module composition $\Omega_\square$ if

$$\Omega_\square : M \times G \longrightarrow M \qquad [5]$$

is a mapping satisfying, for every x, x' ε M and g, g' ε G,

i) $(x + x') \square g = (x \square g) + (x' \square g)$ (left distributivity) [6]

ii) $x \square (g * g') = (x \square g) \square g'$. (associativity) [7]

Similarly, a right G-module M is called a unitary right G-module if

$x \square 1_G = x$, for every $x \in M$. (right neutrality) [8]

(def) Two-sided module over a group

Let G be an abelian group. Then an abelian group M is called a two-sided G-module with a module composition $\Omega_\square$ (which denotes symbolically two mappings for economy of notation) if the mappings

$\Omega_\square : G \times M \longrightarrow M$ (left composition) [9]

and $\Omega_\square : M \times G \longrightarrow M$ (right composition) [10]

make M simultaneously a left G-module (w.r.t. the left composition) and a right G-module (w.r.t. the right composition) satisfying the condition:

$g \square x = x \square g$, [11]

for every $x \in M$ and $g \in G$.

Remark

It is not without reason that G is required to be abelian in defining a two-sided G-module. Consider any $x \varepsilon M$ and $g, g' \varepsilon G$. We have

$$(g * g') \,\square\, x = g \,\square\, (g' \,\square\, x) = g \,\square\, (x \,\square\, g') \qquad \text{(by [11])}$$

$$= (x \,\square\, g') \,\square\, g = x \,\square\, (g' * g) \qquad \text{(by [7])}$$

$$= (g' * g) \,\square\, x \ .$$

To satisfy the above relation for an arbitrary x imposes some conditions on $M, \Omega_\square$ as well as G. The simplest way to achieve this is to require an abelian G. ▮

In the language of "operator group" (cf. [296] and [306] of Chapter 2) the definition of a left G-module can be stated as:

(def) An abelian group M is a left G-module if the group G acts on M. M is a unitary left G-module if G acts on M such that

$$1_G \,\square\, x = x \ ,$$

for every $x \varepsilon M$ (cf. [313] of §.2.17).

Similarly, the right and the two-sided axioms can be stated in a similar manner. Since we use "left" axioms throughout the book (unless otherwise stated), we shall often call a left module simply a "module".

The concept of modules over a ring is similar to that of a group-module.

(def) Left module over a ring

Let R be a ring (written additively and multiplicatively). Then an abelian group M (written also additively, for economy of notation) is called a left R-module (i.e. a left module over the ground ring R) with a module composition $\Omega_\square$ if the mapping

$$\Omega_\square : R \times M \longrightarrow M \qquad [12]$$

satisfies, for each r, r' ε R and x, x' ε M, the following axioms:

i) $r \square (x+x') = (r \square x) + (r \square x')$ (right distributivity) [13]

ii) $(r+r') \square x = (r \square x) + (r' \square x)$ (left distributivity) [14]

iii) $(rr') \square x = r \square (r' \square x)$. (associativity) [15]

If R has 1, and if the additional condition

$1 \square x = x$, for every x ε M, (left neutrality) [16]

is satisfied, then we call M a unitary left R-module (1 is the multiplicative unit of R). A left R-module M is said to be trivial if $r \square x = 0$, for every r ε R, x ε M. Otherwise, M is non-trivial.

Similarly, right R-modules are defined by requiring the mapping

$$\Omega_{\square} : M \times R \longrightarrow M \qquad [17]$$

to satisfy, instead of [13] to [15], the following axioms:

i) $(x+x')\square r = (x\square r)+(x'\square r)$ (right distributivity) [18]

ii) $x\square(r+r') = (x\square r)+(x\square r')$ (left distributivity) [19]

iii) $x\square(rr') = (x\square r)\square r'$. (associativity) [20]

Again, if R has 1 and if the additional condition

$x\square 1 = x$ (right neutrality) [21]

is satisfied, then M is called a unitary right R-module.

(def) Two-sided module over a ring

Let R be an abelian ring with 1. Then an abelian group M is called an R-module (i.e. a module over R) with module compositions $\Omega_{\square}$ if the mappings

$\Omega_{\square} : R \times M \longrightarrow M$ (left composition) [22]

and

$\Omega_{\square} : M \times R \longrightarrow M$ (right composition) [23]

make M simultaneously a left R-module and a right R-module satisfying the condition

$$r \square x = x \square r \,, \qquad [24]$$

for every $r \in R$ and $x \in M$.

Remarks

1) Reader should be able to show that the abelian requirement of the ring R is consistent with [24] (cf. the remark following [11]).

2) There are two very simple but useful properties (we shall state them only for left R-modules, but it is trivial to see that similar properties also hold for right R-modules):

(i) $0' \square x = 0$, for every $x \in M$ [25]

(ii) $r \square 0 = 0$, for every $r \in R$ [26]

(iii) $(-r) \square x = r \square (-x) = -(r \square x)$, [27]

where 0 and 0' are zeros of M and R, respectively. The proofs of [25] to [27] are quite simple. Consider any $x \in M$. Then

$$r \square x = (0' + r) \square x = 0' \square x + r \square x$$

i.e. $0' \square x = 0$. ∎

Next, for every $r \in R$, we have

$$r \square 0 = r \square (0 + 0) = (r \square 0) + (r \square 0)$$

i.e. $$r \square 0 = 0 \text{ .} \quad \blacksquare$$

Finally, we have

$$0 = (r - r) \square x = r \square x + (-r) \square x$$

i.e. $$(-r) \square x = -(r \square x) \text{ .} \qquad [28]$$

On the other hand,

$$0 = r \square (x - x) = r \square x + r \square (-x)$$

leads to

$$r \square (-x) = -(r \square x) \text{ .} \qquad [29]$$

[28] and [29] yield directly [27]. ∎∎

(def) A left vector space is defined as a unitary left module over a field. Similar definition holds for a right vector space.

Examples

1) Let B be a left ideal of a ring R. Then B is a left R- module with ring multiplication as the module composition

since

$$RB \subset B \qquad [30]$$

allows us to define

$$\Omega_{\square} : (r, b) \longmapsto rb \in B, \quad \text{(ring multiplication)} \qquad [31]$$

for every $r \in R$ and $b \in B$.

Similarly, B is a <u>right</u> R-module if B is a right ideal of R.

2) Any abelian group M can be made a left I-module with the module composition defined by

$$\Omega_{\square} : (n, x) \longmapsto nx , \qquad [32]$$

for every $n \in I$ and $x \in M$. As in [29] and [30] of Chapter III, nx is defined by

$$nx = \underbrace{x + \cdots + x}_{n \text{ copies}} , \quad \text{if } n \in I_+ \qquad [33]$$

$$nx = \underbrace{(-x) + \cdots + (-x)}_{|n| \text{ copies}} , \quad \text{if } n \in I_- \qquad [34]$$

and

$$0x = 0 . \qquad [35]$$

3) Let K be a field (written additively and multiplicative
Consider the set, K^n, of all ordered n-tuples defined by

$$\{k_1, \ldots, k_n\}, \; k_i \in K . \qquad [36]$$

Then K^n is an abelian group w.r.t. the addition defined by

$$\{k_1,\ldots, k_n\} + \{k_1',\ldots, k_n'\} = \{k_1 + k_1',\ldots, k_n + k_n'\} \qquad [37]$$

for every k_i, $k_i' \in K$.

Next, we can make K^n a vector space over K (i.e. to introduce a K-module structure into K^n) by defining a module composition $\Omega_\square$:

$$k \square \{k_1, \ldots, k_n\} = \{kk_1, \ldots, kk_n\} , \qquad [38]$$

for every k, $k_i \in K$.

4) Let E be a bounded closed interval in $\mathbb{R}^1$. Then the set Map(E, $\mathbb{R}^1$) is a unitary $\mathbb{R}$-module (i.e. a vector space over $\mathbb{R}$) w.r.t. the composition introduced below. An addition is first defined for Map(E, $\mathbb{R}^1$) by

$$(f + f')x = f(x) + f'(x) , \qquad [39]$$

for every $f, f' \in \mathrm{Map}(E, \mathbb{R}^1)$ and $x \in E$. This makes $\mathrm{Map}(E, \mathbb{R}^1)$ an abelian group. To make it a unitary R-module we introduce the module composition:

$$(a \mathbin{\square} f)x = a(f(x)) \tag{40}$$

for every $a \in \mathbb{R}$. It is necessary to show that [40] is a module composition as we claimed. In other words, we have to show that the following axioms are satisfied:

$$\text{i)} \quad (a + a') \mathbin{\square} f = a \mathbin{\square} f + a' \mathbin{\square} f \tag{41}$$

$$\text{ii)} \quad a \mathbin{\square} (f + f') = a \mathbin{\square} f + a \mathbin{\square} f' \tag{42}$$

$$\text{iii)} \quad (aa') \mathbin{\square} f = a \mathbin{\square} (a' \mathbin{\square} f) \tag{43}$$

$$\text{iv)} \quad 1 \mathbin{\square} f = f , \tag{44}$$

for every $a, a' \in \mathbb{R}^1$ and $f, f' \in \mathrm{Map}(E, \mathbb{R}^1)$. The verification is simple. We shall do the first one and leave the rest to the reader:

$$\begin{aligned} ((a + a') \mathbin{\square} f)x &= (a + a')(f(x)) \\ &= a(f(x)) + a'(f(x)) \\ &= (a \mathbin{\square} f)x + (a' \mathbin{\square} f)x \\ &= (a \mathbin{\square} f + a' \mathbin{\square} f)x \end{aligned}$$

i.e. $\quad (a + a') \mathbin{\square} f = a \mathbin{\square} f + a' \mathbin{\square} f$. ||

§. 4.2. Submodules, quotient modules, faithful module and annihilator

R denotes, in this section, an arbitrary ring.

From now on we shall concentrate only on modules over a ring. Besides, as in previous chapters, we shall use only "left" objects (like left R-module, etc.).

(def) Let M be a left R-module. A sub abelian group M', of M, is called a submodule of M if M' is a left R-module w.r.t. the restriction (to M') of the module composition of M, i.e. if

$$r \,\square\, x' \in M' , \quad \text{for every } r \in R \text{ and } x' \in M' , \qquad [45]$$

where $\Omega_\square$ denotes the module composition on M.

For convenience, we now define the module composition between any subset R' of R and any subset M' of M by

$$R' \,\square\, M' = \{ \sum_{\lambda \in \Lambda} r_\lambda \,\square\, x_\lambda \mid r_\lambda \in R', \ x_\lambda \in M', \ \lambda \in \Lambda ; \ \Lambda = \text{an index set} \qquad [46]$$

In this notation, [45] is equivalent to

$$R \,\square\, M' \subset M' . \qquad [47]$$

(notation) $M' \underset{\text{mod}}{\subset} M$ denotes that M' is a submodule of M.

Similar to the situation with rings, a module M is its own submodule. Hence M is called the improper submodule of itself.

The singleton $\{0\}$ is called the zero submodule of M (we do not use the term "trivial" which has a different meaning, namely if $r \circ x = 0$ for every $r \in R$, $x \in M$). A submodule is a proper one if it is not improper.

(notation) $M' \underset{\text{mod}}{\subsetneq} M$ denotes that M' is a proper submodule of M.

Proposition I

The intersection of any non-empty collection of submodules of a left R-module M is again a submodule of M.

Proof

It is obvious. Let M_1 and M_2 be submodules of M, then, for every $m, m' \in M_1 \cap M_2$ we have

$$m + m' \in M_1 \quad (\text{since } m, m' \in M_1)$$

and $$m + m' \in M_2 \quad (\text{since } m, m' \in M_2)\,.$$

Hence $$m + m' \in M_1 \cap M_2\,.$$

Other axioms for a module are trivially satisfied too.

(def) Quotient modules (i.e. factor modules)

Let M' be a submodule of a left R-module M (with a module composition $\Omega_{\square}$). Then the set

$$M \bmod M' \equiv M/M' \equiv \{\bar{x} \mid \bar{x} = x \bmod M',\ x \in M\} \qquad [48]$$

obviously forms an abelian group w.r.t. the "canonical" addition defined by

$$\bar{x}_1 + \bar{x}_2 \equiv \overline{x_1 + x_2} = (x_1 + x_2) \bmod M' . \qquad [49]$$

We remind the reader that x mod M' is defined by

$$x \bmod M' = \{x + x' \mid x' \in M\}. \qquad [50]$$

A "canonical" module composition $\Omega_{\blacksquare}$ is now introduced:

$$\Omega_{\blacksquare} : R \times (M \bmod M') \longrightarrow M \bmod M' \qquad [51]$$

with $$r \blacksquare \bar{x} = \overline{r \square x} = r \square x \bmod M' . \qquad [52]$$

It is easy to see that M mod M' is a left R-module w.r.t. $\Omega_{\blacksquare}$ (we leave the verification to the reader; or see the Problem Section). The left R-module M mod M' (with $\Omega_{\blacksquare}$) is called the quotient module of M modulo M' .

Remark

If R has 1 so that M is a unitary left R-module then M mod M' is also a unitary left R-module. This is obvious since, for every $x \in M$,

$$1 \blacksquare \bar{x} = \overline{1 \square x} = \bar{x} \,. \qquad [53]$$

(def) A left R-module M (with $\Omega_\square$) is said to be faithful (as the left R-module) if

$$r \in R : r \square M = \{0\} \Longrightarrow r = 0 \,. \qquad [54]$$

(def) The annihilator of a left R-module M (with $\Omega_\square$) is defined as the set

$$\text{ann}.M \equiv \{ a \mid a \in R,\ a \square M = \{0\} \} \,. \qquad [55]$$

Hence the annihilator of a module indicates to what extent the module deviates from being "faithful". It can be easily shown that ann.M is an (two-sided) ideal of R (see the Problem Section). Therefore, we can construct the quotient ring R/ann.M. M can be made a left module over this quotient ring by defining the composition

$$\Omega_\blacksquare : R/\text{ann}.M \times M \longrightarrow M \qquad [56]$$

with $\quad (r \bmod \text{ann.}M) \blacksquare x = r \square x$, [57]

for every $r \in R$ and $x \in M$. It is trivial to verify that [57] is well-defined in the sense:

$$r, r' \in R : (r = r' \mod \text{ann.}M) \Longrightarrow r \square x = r' \square x . \qquad [58]$$

The LHS of [58] can be written as

$$r - r' \in \text{ann.}M \qquad [59]$$

hence $\quad (r - r') \square x = 0$ [60]

i.e. $\quad r \square x = r' \square x.$ [61]

This establishes our claim. It remains to show that [57] satisfies all the module axioms. The verification is again trivial and we leave it to the reader.

§. 4.3. Module homomorphisms

Let R be a ring.

(def) Let M (with $\Omega_{\square}$) and M' (with $\Omega_{\blacksquare}$) be left R-modules. Then a mapping $f \in \text{Map}(M, M')$ is called an R-homomorphism (or R-hom, for short) if

i) $f(x_1 + x_2) = fx_1 + fx_2$ [62]

ii) $f(r \square x) = r \blacksquare (fx)$ [63]

for every $r \in R$ and $x, x_1, x_2 \in M$. We list further:

an R-monomorphism (or R-mono) is an injective R-hom,

an R-epimorphism (or R-epi) is a surjective R-hom,

an R-isomorphism (or R-iso) is a bijective R-hom,

an R-endomorphism (or R-endo) is an R-hom whose codomain coincides with its domain,

an R-automorphism is a bijective R-endo.

(notation)

$\mathrm{Hom}_R(M, M') \equiv$ the set of all R-homomorphisms from M to M' (this refers to their R-module structures).

$\mathrm{Hom}(M, M') \equiv$ the set of all abelian-group homomorphisms from M to M' (this refers to their abelian group structures only).

Remark

An R-homomorphism is often referred to, by some authors, as an "R-linear mapping".

Similarly, we shall use the notation:

$$\mathrm{Mon}_R(M, M'), \quad \mathrm{Epi}_R(M, M'), \quad \mathrm{Iso}_R(M, M'), \quad \mathrm{End}_R M, \text{ etc.}$$

and $\quad \mathrm{Mon}(M, M'), \mathrm{Epi}(M, M'), \mathrm{Iso}(M, M'), \mathrm{End}.M$, etc.

We have already shown (Proposition V, Chapter II) that Hom(M, M') is an abelian group whose group composition was defined by [22] of Chapter II. It is easy to see that $\mathrm{Hom}_R(M, M')$ is a sub abelian group of Hom(M, M'), and we leave it to the reader to verify (or see the Problem Section).

<u>Proposition II</u>

For a left R-module M (with $\Omega_\square$), $\mathrm{End}_R M$ is a subring of End.M. The <u>additive and multiplicative compositions</u> of the <u>ring</u> structure are defined by [21] and [22] of Chapter III.

<u>Proof</u>

In Problem 6 in the Problem Section of Chapter III, we have shown that End.M forms a <u>ring</u> with 1 (the multiplicative unit is just the <u>identity</u> endomorphism). Therefore, since $\mathrm{End}_R M$ is a subset of End.M, we have only to check the "closure" property. For every $r \in R$, $x \in M$ and $f_1, f_2 \in \mathrm{End}_R M$, we have

$$(f_1 \circ f_2)(r \square x) = f_1(f_2(r \square x)) = f_1(r \square (f_2 x))$$

$$= r \square (f_1(f_2 x)) = r \square ((f_1 \circ f_2)x) \qquad [64]$$

i.e. $$f_1 \circ f_2 \in \mathrm{End}_R M \ . \qquad [65]$$

Hence the closure property is proved. ‖

(def) Let M' be a submodule of a left R-module M, then the mapping

$$\eta : M \longrightarrow M/M' \qquad [66]$$

with $\eta : x \longmapsto \eta x \equiv x \bmod M'$, for every $x \in M$, [67]

is called the canonical R-homomorphism from M to M/M'. To see that [67] is indeed an R-hom we now compute, for every $x_1, x_2 \in M$ and $r \in R$,

$$\begin{aligned} \eta(x_1 + x_2) &= (x_1 + x_2) \bmod M' \\ &= x_1 \bmod M' + x_2 \bmod M' \\ &= \eta x_1 + \eta x_2 \end{aligned} \qquad [68]$$

and
$$\begin{aligned} \eta(r \square x) &= r \square x \bmod M' \\ &= r \blacksquare (x \bmod M') \qquad (\text{cf.}[52]) \\ &= r \blacksquare (\eta x) \ . \end{aligned} \qquad [69]$$

Remark

The canonical R-homomorphism η is obviously an

R-epimorphism.

Proposition III

Let M and P be left R-modules and $f \in \mathrm{Hom}_R(M, P)$. If M' and P' are, respectively, submodules of M and P such that

$$fM' \subset P' \tag{70}$$

then the following diagram is commutative:

$$\begin{array}{ccc} M & \xrightarrow{f} & P \\ {\scriptstyle \eta_1}\big\downarrow & & \big\downarrow{\scriptstyle \eta_2} \\ M/M' & \xrightarrow{f^*} & P/P' \end{array} \tag{71}$$

where η_1 and η_2 are canonical R-homomorphisms, and f^* is defined by

$$f^* : M/M' \longrightarrow P/P' \tag{72}$$

with $$f^* : x \bmod M' \longmapsto fx \bmod P' , \tag{73}$$

for every $x \in M$.

Proof

It is trivial since, for every $x \in M$, we have

$$(\eta_2 \circ f)x = \eta_2 (fx) = fx \bmod P' \qquad [74]$$

and $$(f^* \circ \eta_1)x = f^*(\eta_1 x) = f^*(x \bmod M') = fx \bmod P' \qquad [75]$$

i.e. $$(\eta_2 \circ f)x = (f^* \circ \eta_1)x \, . \qquad [76]$$

Hence $$\eta_2 \circ f = f^* \circ \eta_1 \, . \quad \| \qquad [77]$$

(def) Kernel and cokernel of an R-hom

The definition of the kernel of an R-homomorphism is similar to that of a ring-homomorphism; it is the preimage of zero. Let M, M' be left R-modules and $f \in \mathrm{Hom}_R(M, M')$ then it is easy to see that

$$\mathrm{ker}.f \underset{\mathrm{mod}}{\subset} M \qquad [78]$$

and $$\mathrm{im}.f \underset{\mathrm{mod}}{\subset} M' \, . \qquad [79]$$

Therefore we can form the quotient modules

$$M/\mathrm{ker}.f \equiv \mathrm{coim}.f$$

and $$M'/\mathrm{im}.f \equiv \mathrm{coker}.f, \qquad [80]$$

to be called coimage and cokernel of f. We note that the R-iso

$$\text{coim.f} \longleftrightarrow \text{im.f} \qquad [81]$$

follows immediately from the next Proposition. For this reason there is actually no urgent need for introducing coim.f which is listed here only to complete the "dual" symmetry.

It is obvious that if f is an R-hom, then

$$\text{ker.f} = \{0\} \Longleftrightarrow \text{f is an } \underline{\text{R-monomorphism.}} \qquad [82]$$

Proposition IV

Let M and P be left R-modules, then

$$f \in \text{Epi}_R(M, P) \Longrightarrow P \longleftrightarrow M/\text{ker.f} . \qquad [83]$$

Proof

For convenience let us write

$$\overline{x} \equiv x \text{ mod ker.f} , \ x \in M . \qquad [84]$$

Consider the mapping

$$\alpha : M/\text{ker.f} \longrightarrow P$$

with $\alpha : \overline{x} \longmapsto fx$, for every $x \in M$. [85]

α is clearly <u>surjective</u> since f is. If x_1 and x_2 are elements such that

$$fx_1 = fx_2$$

then $$f(x_1 - x_2) = 0$$

i.e. $$x_1 - x_2 \in \ker.f.$$

Hence $$\bar{x}_1 = \bar{x}_2 \qquad [86]$$

i.e. α is <u>injective</u>. Therefore, α is <u>bijective.</u>

It is straightforward to see that α is an R-hom, since, for every $r \in R$, we have

$$\alpha(\bar{x}_1 + \bar{x}_2) = \alpha(\overline{x_1 + x_2}) = f(x_1 + x_2)$$

$$= fx_1 + fx_2 = \alpha\bar{x}_1 + \alpha\bar{x}_2 \qquad [87]$$

and $$\alpha(r \blacksquare \bar{x}) = \alpha(\overline{r \square x}) = f(r \square x) = r \square (fx) = r \square (\alpha\bar{x}), \qquad [88]$$

where we have used the notation of [52]. Besides, M and P share the same notation for their different module compositions. ∎

§. 4.4. Sum and direct sum of modules

Let M be a left R-module, written additively as an abelian group, with a module composition $\Omega_{\square}$ (unless otherwise specified).

(def) Let M_1 and M_2 be submodules of M, then the module sum of M_1 and M_2 is defined by:

$$M_1 + M_2 \equiv \{m \mid m = m_1 + m_2,\ m_1 \in M_1,\ m_2 \in M_2\}. \qquad [89]$$

For enconomy of language we shall call a module sum simply a sum when there is no possibility of ambiguity.

Clearly,

$$(M_1 + M_2) \underset{\text{mod}}{\subset} M. \qquad [90]$$

This is trivial since any $m, m' \in (M_1 + M_2)$ can be written as

$$m \equiv m_1 + m_2 \quad , \quad m' \equiv m_1' + m_2' \qquad [91]$$

with $m_1, m_1' \in M_1$ and $m_2, m_2' \in M_2$. Therefore,

$$m + m' = (m_1 + m_2) + (m_1' + m_2')$$

$$= (m_1 + m_1') + (m_2 + m_2') \; \varepsilon \; (M_1 + M_2) \; . \quad \blacksquare$$

We note that $M_1 + M_2$ is the <u>smallest submodule</u> containing both M_1 and M_2. This is equivalent to the following statement: for any <u>submodules</u> $M_i (i = 1, 2, 3)$ of M,

$$M_3 \underset{\text{set}}{\supset} (M_1 \cup M_2) \Longrightarrow M_3 \underset{\text{mod}}{\supset} (M_1 + M_2) \; . \qquad [92]$$

This is obvious since, by the LHS of [92],

$$m_1 \; \varepsilon \; M_1 \Longrightarrow m_1 \; \varepsilon \; M_3$$

and $$m_2 \; \varepsilon \; M_2 \Longrightarrow m_2 \; \varepsilon \; M_3 \; .$$

Thus $$(m_1 + m_2) \; \varepsilon \; M_3$$

i.e. $$(M_1 + M_2) \underset{\text{mod}}{\subset} M_3 \; . \quad \blacksquare$$

The definition [89] can obviously be generalized to the case of more than two submodules. In general we define

$$\sum_{\lambda \varepsilon \Lambda} M_\lambda \equiv \{ m \mid m = \sum_{\lambda \varepsilon \Lambda} m_\lambda \; , \; m_\lambda \; \varepsilon \; M_\lambda , \; m_\lambda = 0 \text{ p.p. } \lambda \varepsilon \Lambda \} \qquad [93]$$

where Λ is any index set, and

$$m_\lambda = 0 \text{ p.p. } \lambda \in \Lambda \qquad [94]$$

means that only <u>finitely many</u> m_λ are not zero. This is often expressed as "$m_\lambda = 0$ <u>for almost all</u> $\lambda \in \Lambda$" (p.p. = the abbreviation of "presque partout" in French = "for almost all").

<u>Proposition V</u> (Dedekind's law)

Let M_1, M_2, M_3 be submodules of a left R-module M with either $M_1 \subset M_3$ or $M_2 \subset M_3$ (or both) then

$$(M_1 + M_2) \cap M_3 = M_1 \cap M_3 + M_2 \cap M_3 \quad . \qquad [95]$$

<u>Proof</u>

We shall show [95] by establishing the inclusion relation in both directions.

First it is clear that

$$\text{LHS of } [95] \supset \text{ RHS of } [95]$$

since every element of the RHS of [95] has the form

$$m_{13} + m_{23} \quad \text{with} \quad m_{13} \in M_1 \cap M_3 \quad \text{and} \quad m_{23} \in M_2 \cap M_3 \ .$$

Hence

$$m_{13} + m_{23} \in M_1 + M_2$$

and $m_{13} + m_{23} \varepsilon M_3$

i.e. $m_{13} + m_{23} \varepsilon (M_1 + M_2) \cap M_3 = \text{LHS of } [95]$. ▌

To show that

$$\text{LHS of } [95] \subset \text{RHS of } [95], \qquad [96]$$

let us consider any $x \varepsilon (M_1 + M_2) \cap M_3$. Then x has the form

$$x \equiv m_1 + m_2 \quad \text{with} \quad m_1 \varepsilon M_1 \text{ and } m_2 \varepsilon M_2 \ . \qquad [97]$$

If $M_1 \subset M_3$, then [97] implies

$$m_2 = (x - m_1) \varepsilon (M_3 + M_1) \varepsilon M_3$$

i.e. $m_2 \varepsilon M_2 \cap M_3$

Hence $x = (m_1 + m_2) \varepsilon M_1 + M_2 \cap M_3 = M_1 \cap M_3 + M_2 \cap M_3$

which establishes [96]. On the other hand, if $M_2 \subset M_3$, then by symmetry (i.e. exchange M_1 with M_2) we also reach [96]. This completes the proof. ▌▌

<u>Remark</u>

Effectively [95] has the form

$$(M_1 + M_2) \cap M_3 = M_1 + M_2 \cap M_3 \text{ , if } M_1 \subset M_3$$

and the form

$$(M_1 + M_2) \cap M_3 = M_1 \cap M_3 + M_2 \quad \text{if } M_2 \subset M_3 . \blacksquare$$

(def) Internal direct sum

Let $\{M_\lambda\}_{\lambda \varepsilon \Lambda}$ be a collection of submodules of M such that

i) $$M = \sum_{\lambda \varepsilon \Lambda} M_\lambda \qquad [98]$$

ii) For any $m \varepsilon M$, the decomposition

$$m = \sum_{\lambda \varepsilon \Lambda} m_\lambda, \text{ with } m_\lambda \varepsilon M_\lambda , \qquad [99]$$

is unique. Then M is called the (internal) direct sum of the submodules M_λ, $\lambda \varepsilon \Lambda$. Each M_μ is called a direct summand of M, or a direct supplement to the module

$$\sum_{\substack{\lambda \varepsilon \Lambda \\ \lambda \neq \mu}} M_\lambda .$$

(notation) The internal direct sum M with direct summands M_λ is written as (an ordered sum)

$$M = \bigoplus_{\lambda \varepsilon \Lambda} M_\lambda . \qquad [100]$$

Or, sometimes, for a finite number of direct summands we write

$$M = M_1 \oplus \cdots \oplus M_n . \qquad [101]$$

<u>Proposition VI</u>

Let $\{M_\lambda\}_{\lambda \varepsilon \Lambda}$ be a collection of submodules of a left R-module M, then the following conditions are equivalent:

i) $M = \bigoplus_{\lambda \varepsilon \Lambda} M_\lambda .$ [102]

ii) $M_\mu \cap \sum_{\substack{\lambda \varepsilon \Lambda \\ \lambda \neq \mu}} M_\lambda = \{0\}$, for every $\mu \varepsilon \Lambda$. [103]

iii) $\sum_{\lambda \varepsilon \Lambda} m_\lambda = 0, \; m_\lambda \varepsilon M_\lambda \Longrightarrow m_\lambda = 0$, for every $\lambda \varepsilon \Lambda$. [104]

<u>Proof</u>

We shall proceed with the proof in the order:

$$\text{(i)} \Longrightarrow \text{(ii)} \Longrightarrow \text{(iii)} \Longrightarrow \text{(i)} . \qquad [105]$$

<u>Step 1</u> (i) $\Longrightarrow$ (ii).

Let non-zero $m_\mu \varepsilon M_\mu$ then [102] implies that the only <u>non-zero</u> component of m_μ under the decomposition [99] is the

μth component. On the other hand, the μth component of an element of

$$\sum_{\substack{\lambda \in \Lambda \\ \lambda \neq \mu}} M_\lambda \qquad [106]$$

must be zero since $\lambda \neq \mu$ is imposed on the summation [106]. Hence

$$\text{non-zero } m_\mu \notin \sum_{\substack{\lambda \in \Lambda \\ \lambda \neq \mu}} M_\lambda \qquad [107]$$

i.e.

$$M_\mu \cap \sum_{\substack{\lambda \in \Lambda \\ \lambda \neq \mu}} M_\lambda = \{0\}. \qquad [108]$$

In other words, [103] is established. ▌

<u>Step 2</u> (ii) $\Longrightarrow$ (iii).

Let us prove this by the method of contradiction. Assume that

$$\sum_{\lambda \in \Lambda} m_\lambda = 0 \qquad [109]$$

with at least some $m_\mu \neq 0$. We have, by [109],

$$m_\mu + \sum_{\substack{\lambda \in \Lambda \\ \lambda \neq \mu}} m_\lambda = 0 \qquad [110]$$

i.e. (non-zero) $m_\mu = \sum_{\substack{\lambda \in \Lambda \\ \lambda \neq \mu}} - m_\lambda \in \sum_{\substack{\lambda \in \Lambda \\ \lambda \neq \mu}} M_\lambda$ [111]

or $M_\mu \cap \sum_{\substack{\lambda \in \Lambda \\ \lambda \neq \mu}} M_\lambda \neq \{0\}$ [112]

which contradicts [103]. Consequently, $m_\mu = 0$. ∎

Step 3 (iii) $\Longrightarrow$ (i).

Assume that, for some $m \in M$, there are two possible decompositions :

$$m = \sum_{\lambda \in \Lambda} m_\lambda \quad \text{and} \quad m = \sum_{\lambda \in \Lambda} m_\lambda' . \qquad [113]$$

Subtraction of one from the other yields

$$0 = \sum_{\lambda \in \Lambda} (m_\lambda - m_\lambda') . \qquad [114]$$

But [104] requires

$$(m_\lambda - m_\lambda') = 0 \text{ , for every } \lambda \in \Lambda \qquad [115]$$

i.e. $m_\lambda = m_\lambda'$.

In other words, the decomposition of any $m \in M$ is unique. ||

Remark

The equivalence of [102] and [103] for a sum leads to the following statement:

Let M_1 and M_2 be submodules of M, then the sum $M_1 + M_2$ is a direct sum iff

$$M_1 \cap M_2 = \{0\} . \qquad [116]$$

(def) External direct sum

Let $\{M_i\}_{i=1,\ldots,n}$ be a set of R-modules. Denote by

$$M_1 \dot{+} \cdots \dot{+} M_n \quad \text{or} \quad \dot{\sum_{i=1}^{n}} M_i \qquad [117]$$

the set of all n-tuples,

$$\{m_1, \ldots, m_n\} , \quad m_i \in M_i . \qquad [118]$$

Then $\dot{\sum_{i=1}^{n}} M_i$ is clearly a left R-module w.r.t. the composition

$$\Omega_{\blacksquare} : R \times \dot{\sum_{i=1}^{n}} M_i \longrightarrow \dot{\sum_{i=1}^{n}} M_i \qquad [119]$$

with $$r \blacksquare \{m_1, \ldots, m_n\} = \{r \square m_1, \ldots, r \square m_n\},$$ [120]

for every $m_i \varepsilon M_i$ and $r \varepsilon R$. The left module $\dot{\sum}_{i=1}^{n} M_i$ is called the (external) direct sum of M_i, i=1, ..., n .

Remark

An external direct sum $\dot{\sum}_{i=1}^{n} M_i$ can be written in the form of an internal direct sum by introducing:

$$M_i' \equiv \{m \mid m = \{0, \ldots, 0, m_i, 0, \ldots, 0\}, m_i \varepsilon M_i\}$$ [121]

where m_i occupies the ith position of the n-tuple. Clearly, we have

$$\dot{\sum_{i=1}^{n}} M_i = \bigoplus_{i=1}^{n} M_i' .$$ [122]

Proposition VII

Let M_1 and M_2 be submodules of M such that

$$M = M_1 + M_2 \quad \text{(not a direct sum)}$$ [123]

then, for any submodule M_3 (of M) containing M_1,

$$M_3 = M_1 + (M_2 \cap M_3) \ . \qquad [124]$$

<u>Proof</u>

For any $m_3 \varepsilon M_3$ we can write, according to [123],

$$m_3 = m_1 + m_2 \quad \text{with} \quad m_i \varepsilon M_i \qquad [125]$$

i.e. $$m_2 = m_3 - m_1 \varepsilon M_3 \ .$$

Hence $$m_2 \varepsilon M_2 \cap M_3 \qquad [126]$$

i.e. $$M_3 = M_1 + (M_2 \cap M_3) \ . \ \|$$

<u>Proposition VIII</u>

Let M_1 and M_2 be submodules of M then

$$(M_1 + M_2)/M_1 \longleftrightarrow M_2/(M_1 \cap M_2) \ . \qquad [127]$$

<u>Proof</u>

First, we note that every element in the quotient module $(M_1 + M_2)/M_1$ is of the form, <u>set-theoretically</u>,

$$m_2 \bmod M_1 \ , \quad \text{with} \quad m_2 \varepsilon M_2 \ . \qquad [128]$$

But in order to write [128] as a <u>quotient module</u> we have to replace it by $(M_1 + M_2) \bmod M_1$. This is because M_1 is not necessarily contained in M_2, but $M_1 + M_2$ is always a submodul of M_2 (cf. the definition of a quotient module). The proof is then trivial. The required isomorphism is defined by

$$f : (M_1 + M_2)/M_1 \longrightarrow M_2/(M_1 \cap M_2) \qquad [129]$$

with

$$f : m_2 \bmod M_1 \longmapsto m_2 \bmod (M_1 \cap M_2), \qquad [130]$$

for every $m_2 \in M_2$.

That f is an R-isomorphism is not difficult to see. First, f is clearly an R-homomorphism and it is surjective. Next, f is injective since (by inspection of the RHS of [130])

$$\begin{aligned} \ker.f &= \{m_2 \bmod M_1 \mid m_2 \in M_2,\ m_2 \in (M_1 \cap M_2)\} \\ &= \{m_2 \bmod M_1 \mid m_2 \in M_1 \cap M_2\} \\ &= \{0 \bmod M_1\}_1 \ . \end{aligned} \qquad [131]$$

Consequently f is an R-iso. (We remind the reader that the subscript of $\{\ \}_1$ emphasizes the set is a singleton) ||

The following proposition is the module-theoretical version of the Zassenhaus lemma. The group-theoretical version of the lemma was discussed in Proposition XVII of of Chapter II (cf. especially [141], of Chapter II).

<u>Proposition IX</u> (Zassenhaus lemma)

Let M_1, M_2, M_3, M_4 be submodules of M with

$$M_1 \subset M_3 \quad \text{and} \quad M_2 \subset M_4 \tag{132}$$

then

$$(M_1 + M_3 \cap M_4)/(M_1 + M_2 \cap M_3) \longleftrightarrow (M_2 + M_3 \cap M_4)/(M_2 + M_1 \cap M_4) . \tag{133}$$

<u>Proof</u>

Use Proposition VIII, but make the following substitutions in [127]:

$$M_1 \quad \text{by} \quad M_1 + M_2 \cap M_3 \tag{134}$$

and

$$M_2 \quad \text{by} \quad M_3 \cap M_4 . \tag{135}$$

Then [127] becomes:

$$((M_1 + M_2 \cap M_3) + M_3 \cap M_4)/(M_1 + M_2 \cap M_3) \longleftrightarrow (M_3 \cap M_4)/((M_1 + M_2 \cap M_3) \cap (M_3 \cap M_4)) . \tag{136}$$

But, set-theoretically, we have

$$(M_1 + M_2 \cap M_3) + M_3 \cap M_4 = M_1 + (M_3 \cap M_2 + M_3 \cap M_4)$$
$$= M_1 + M_3 \cap M_4 \qquad [137]$$

and, by Dedekind's law (Proposition V),

$$(M_1 + M_2 \cap M_3) \cap (M_3 \cap M_4)$$
$$= M_1 \cap (M_3 \cap M_4) + (M_2 \cap M_3) \cap (M_3 \cap M_4)$$
$$= M_1 \cap M_4 + M_2 \cap M_3 \qquad [138]$$

where the fact $(M_2 \cap M_3) \subset (M_3 \cap M_4)$ was used in the application of Dedekind's law.

By substituting [137] into the numerator of the LHS of [136] and [138] into the denominator of the RHS of [136], we obtain

$$(M_1 + M_3 \cap M_4)/(M_1 + M_2 \cap M_3) \longleftrightarrow$$
$$(M_3 \cap M_4)/(M_1 \cap M_4 + M_2 \cap M_3) \ . \qquad [139]$$

Similarly, by symmetry (cf. [167] of §.2.13 of Chapter II), we have (by exchanging the subscripts: 1 with 2 and 3 with 4, in [139]):

$$(M_2 + M_4 \cap M_3)/(M_2 + M_1 \cap M_4) \longleftrightarrow$$

$$(M_4 \cap M_3)/(M_2 \cap M_3 + M_1 \cap M_4) \; . \qquad [140]$$

But the RHS of [139] and [140] are identical, hence their LHS ar isomorphic, i.e.

$$(M_1 + M_3 \cap M_4)/(M_1 + M_2 \cap M_3) \longleftrightarrow$$

$$(M_2 + M_3 \cap M_4)/(M_2 + M_1 \cap M_4) \; . \quad \blacksquare$$

§. 4.5. Chain condition, Artin and Noether modules, maximum and minimum conditions.

The concept of chain conditions plays an important role in modules just as in ring or group theory. In fact, a number of properties that hold in ring theory can be translated into modul without much labor. We shall first lay down the corresponding definitions. Again, in this section, R denotes a ring.

(def) An ascending chain of submodules of a left R-module M is a sequence of distinct submodules (of M),

$$M_1, M_2, \ldots \qquad [141]$$

with

$$M_1 \underset{\text{mod}}{\subset} M_2 \underset{\text{mod}}{\subset} \cdots \; . \qquad [142]$$

The quotient modules M_{i+1}/M_i are called the chain quotients. As in the group theory or ring theory, the chain is said to have a finite length n if the chain

$$M_1 \underset{\text{mod}}{\subset} M_2 \underset{\text{mod}}{\subset} \cdots \underset{\text{mod}}{\subset} M_n \equiv M \qquad [143]$$

terminates at M and has n members in the sequence.

Similarly, we define:

(def) A descending chain of submodules of a left R-module M is a sequence of distinct submodules (of M),

$$M_1, M_2, \ldots \qquad [144]$$

with

$$M_1 \underset{\text{mod}}{\supset} M_2 \underset{\text{mod}}{\supset} \cdots . \qquad [145]$$

The quotient modules M_i/M_{i+1} are called the chain quotients. A descending chain has a length n if

$$M_1 \underset{\text{mod}}{\supset} M_2 \underset{\text{mod}}{\supset} \cdots \underset{\text{mod}}{\supset} M_n \equiv \{0\} . \qquad [146]$$

(def) A left R-module M is said to satisfy the ascending chain condition (or "ACC" for short) if every ascending chain of submodules (of M) has a finite length.

(def) M is said to satisfy the descending chain condition (or "DCC" for short) if every descending chain of submodules (of M) has a finite length.

Just as in ring theory, "Artin" and "Noether" modules are defined:

(def) A left R-module is said to be Noetherian if it satisfies the ACC.

(def) A left R-module is said to be Artinian if it satisfies the DCC.

Remark

It is therefore obvious that a submodule of a Noetherian left R-module is also Noetherian. Similarly, a submodule of an Artinian left R-module is also Artinian.

We now introduce the concept of maximum and minimum conditions:

(def) A proper submodule M', of a left R-module M, is called a maximal submodule (of M) if M' is not properly contained in any proper submodule of M.

(def) A non-zero submodule M', of a left R-module M, is called a minimal submodule (of M) if M' does not properly contain any non-zero submodule of M.

(def) A left R-module M is said to satisfy the maximum condition if, under the partial ordering by set-inclusion, every non-empty collection of submodules (of M) has a maximal member (i.e. a maximal submodule of the collection).

(def) A left R-module M is said to satisfy the minimum condition if, under the partial ordering by set-inclusion, every non-empty collection of submodules (of M) has a minimal member (i.e. a minimal submodule of the collection).

Now we want to show the equivalence between the DCC and the minimum condition. To show this common feature between ring theory and modules we take Proposition XVI of §.3.9. and translate it into module language (the reader should be convinced that this is a waste of time. If so, he catches the real spirit of "homological algebra").

Proposition X

The minimum condition is equivalent to the DCC.

Proof

Let M be a left R-module satisfying the minimum condition. To show that M satisfies the DCC let us consider any descending chain of submodules of M:

$$M_1 \underset{\text{mod}}{\supset} M_2 \underset{\text{mod}}{\supset} \cdots \qquad [147]$$

then the collection $\{M_i\}_{i=1,2,\ldots}$ has a minimal member, say, M_n, since M satisfies the minimum condition. But

$$M_i \subset M_n, \quad \text{for every} \quad i \geq n, \tag{148}$$

hence the chain must break at $i = n$. In other words, M satisfies the DCC.

To show the converse let us assume now that M satisfies the DCC. Consider any non-empty collection, X, of submodules of M. Let X be partially ordered by set-inclusion. Then, for any $x \in X$, x is either a minimal submodule (in X) or not. If x is minimal then the proof is done. If x is not minimal, then

$$\exists\, x_1 (\neq x) \in X : x_1 \subset x . \tag{149}$$

Again x_1 is either minimal or not. If x_1 is minimal then the proof stops here. If not, then

$$\exists\, x_2 (\neq x_1) \in X : x_2 \subset x_1 . \tag{150}$$

By repeating this process we get a descending chain

$$x \underset{\text{mod}}{\supset} x_1 \underset{\text{mod}}{\supset} x_2 \underset{\text{mod}}{\supset} \cdots \tag{151}$$

which, however, must terminate at a finite length since M satisfies the DCC. The last member of the finite chain is then a minimal submodule in X. ||

Similarly, we have the following theorem.

Proposition XI

The maximum condition is equivalent to the ACC.

Proof

It is similar to that of the preceding proposition and we leave this to the reader. ||

Remark

As far as the chain conditions are concerned, a module may satisfy the ACC without satisfying the DCC and vice versa. The well-known examples are:

i) The set I (of all integers) can be considered as an I-module w.r.t. the module composition

$$\Omega_{\square}: I \times I \longrightarrow I, \quad \text{with} \qquad [152]$$

$$\Omega_{\square}: (a, b) \longmapsto ab \quad \text{(ordinary number multiplication)}, \qquad [153]$$

for every a, b ε I. Then I (as an I-module) satisfies the ACC but not the DCC. (We leave the proof to the reader; or see the Problem Section)

ii) Let p be a fixed prime number. Consider the set

$$R_p \equiv \{x/y \mid x,\ y \in I,\ y \text{ is non-zero and indivisible by } p\} \qquad [154]$$

then R_p is a sub additive group of $\mathbb{Q}$ (considered as an additive group). Therefore we can form the <u>quotient</u> (additive) <u>group</u>

$$\mathbb{Q} \bmod R_p \equiv \mathbb{Q}_p. \qquad [155]$$

$\mathbb{Q}_p$ can be considered now as an <u>I-module</u> by defining the module composition,

$$\Omega_\square :\ I \times \mathbb{Q}_p \longrightarrow \mathbb{Q}_p\ , \qquad [156]$$

with $\Omega_\square : (i,\ q \bmod R_p) \longmapsto iq \bmod R_p\ ,$ [157]

for every $i \in I$ and $q \in \mathbb{Q}$. Then the I-module $\mathbb{Q}_p$ satisfies the DCC but not the ACC. (We leave the proof to the reader; or see the Problem Section)

<u>Proposition XII</u>

Let M' be a submodule of a left R-module, then M is <u>Noetherian</u> iff both M' and M/M' are <u>Noetherian</u>. The above statement is also true if the words "Noetherian" are replaced by "Artinian".

<u>Proof</u>

<u>The necessity proof</u>. Assume that M is <u>Noetherian</u>, i.e. it

satisfies the ACC. Clearly M', as a submodule, also satisfies the ACC. By the canonical R-hom, each submodule of M is mapped to a submodule of M/M' by "passing to the quotient". Since the canonical R-hom is surjective, M/M' satisfies the ACC. ∎

The sufficiency proof. We now assume that both M' and M/M' are Noetherian, i.e. they satisfy the ACC. For any ascending chain Σ , of submodules of M,

$$\Sigma : \quad M_1 \subset M_2 \subset \cdots \tag{158}$$

let us construct the following ascending chains

$$\Sigma' : \quad (M_1 + M')/M' \subset (M_2 + M')/M' \subset \cdots \tag{159}$$

$$\Sigma_+ : \quad (M_1 + M') \subset (M_2 + M') \subset \cdots \tag{160}$$

$$\Sigma_\cap : \quad (M_1 \cap M') \subset (M_2 \cap M') \subset \cdots . \tag{161}$$

It is trivial that Σ_+ is in a 1-1 correspondence with Σ' . Since Σ' is an ascending chain of submodules of M/M' which satisfies the ACC by assumption, therefore, Σ_+ has a finite length. Meanwhile the fact

$$M_i \cap M' \underset{\text{mod}}{\subset} M' \tag{162}$$

implies that $\Sigma_\cap$ is also an ascending chain of submodules of M'. Consequently $\Sigma_\cap$ has a <u>finite</u> length since M' satisfies the ACC.

Now we want to show that

(both Σ_+ and $\Sigma_\cap$ have finite lengths) $\Longrightarrow$ (M satisfies the ACC).

The proof is rather simple. Denote by m and n the (finite) lengths of Σ_+ and $\Sigma_\cap$. Then the <u>finiteness</u> of the chains implies that

$$M_i + M' = M_m + M', \text{ for every } i \geq m \qquad [163]$$

and $$M_j \cap M' = M_n \cap M', \text{ for every } j \geq n. \qquad [164]$$

There are two possibilities: $m \geq n$ or $n \geq m$. Consider now the case of $m \geq n$. We have, for $i \geq m$,

$$\begin{aligned} M_i &= M_i \cap (M_i + M') && \text{(set-theoretical)} \\ &= M_i \cap (M_m + M') && \text{(by [163])} \\ &= M_i \cap M_m + M_i \cap M' && \text{(Dedekind's law)} \\ &= M_m + M_i \cap M' && (M_m \subset M_i \text{ for } i \geq m) \\ &= M_m + M_m \cap M' && \text{(by [164] and } m \geq n) \\ &= M_m . && \text{(set-theoretical)} \qquad [16 \end{aligned}$$

Hence the ascending chain

$$M_1 \subset M_2 \subset \cdots \qquad [166]$$

stops at a finite length m. The case of $n > m$ leads to the finite ascending chain

$$M_1 \subset M_2 \subset \cdots \subset M_n . \qquad [167]$$

(We leave the proof of this case to the reader. Or see the Problem Section) Hence the chain [158] has a finite length, i.e. M satisfies the ACC. This completes the proof of the Noetherian statement of the Proposition. (The proof for the Artinian statement can be carried out in exactly the same way, therefore it will not be repeated here) ||

Proposition XIII

If a left R-module M is expressible as a finite sum of Noetherian submodules (of M) then M is also Noetherian.

Proof

By the assumption of the Proposition, we can write

$$M = \sum_{i=1}^{n} M_i \qquad [168]$$

where M_i are submodules (of M) satisfying the ACC. We shall carry out the proof by mathematical induction on the positive integer n. Assume, by induction, that the Proposition is true for a sum of n-1 Noetherian submodules. In other words we now assume that

$$\sum_{i=1}^{n-1} M_i \equiv M', \text{ say} \tag{169}$$

satisfies the ACC. We have to show that

$$\sum_{i=1}^{n} M_i = M$$

also satisfies the ACC (hence Noetherian). First, we have

$$\begin{aligned} M \bmod M' &= M' + M_n \bmod M' \\ &\longleftrightarrow M_n \bmod M' \\ &\longleftrightarrow M_n \bmod (M' \cap M_n) \ . \end{aligned} \tag{170}$$

Since M' satisfies the ACC, $M_n \bmod (M' \cap M_n)$ also satisfie[s] the ACC in virtue of Proposition XII. By [170], we see that M mod M' too satisfies the ACC. The fact that both M' and M mod M' satisfy the ACC implies, by Proposition XII (its "sufficiency" statement), that M satisfies the ACC. This completes the induction. ||

<u>Proposition XIV</u>

If a left R-module M is expressible as a finite sum of

Artinian submodules (of M) then M is also Artinian.

Proof

The proof is similar to that of the preceding theorem therefore we shall not repeat the arguments here. ||

§. 4.6. Submodule generated by a subset and R-linear independence.

In this section R denotes an arbitrary ring, unless otherwise specified. M denotes a left R-module (unless otherwise specified), its module composition will be denoted by $\Omega_{\square}$. Whenever M is assumed to be unitary, the unitarity of R is understood.

(def) Let H be a subset of a left R-module M, then the smallest submodule (of M) containing H is called the submodule (of M) generated by H. A submodule generated by a single element (of M) is called a principal module.

(notation) Mod((H)) = the submodule generated by H.

Remark

We note the notational difference between "Mod (( ))" and "mod". The latter has the meaning of "modulo".

It is clear, by nature of its construction, that the smallest submodule containing H is given by the set

$$\left\{ \sum_{\lambda \in \Lambda} r_\lambda \square h_\lambda + \sum_{\lambda \in \Lambda} i_\lambda h_\lambda \;\middle|\; h_\lambda \in H,\ r_\lambda \in R,\ r_\lambda = 0 \text{ p.p. } \lambda \in \Lambda,\ i_\lambda \in I \right\} \quad [171]$$

where $i_\lambda h_\lambda \equiv h_\lambda + \cdots + h_\lambda$ (i_λ times, cf. [33] to [35]).

In the case of a unitary R (i.e. with 1) and a unitary M, we have

$$\begin{aligned} i_\lambda h_\lambda &= h_\lambda + \cdots + h_\lambda \qquad (i_\lambda \text{ times}) \\ &= 1 \square h_\lambda + \cdots + 1 \square h_\lambda \\ &= (1 + \cdots + 1) \square h_\lambda \\ &= (i_\lambda 1) \square h_\lambda \end{aligned}$$

where $i_\lambda 1 \equiv 1 + \cdots + 1$ (i_λ copies of 1, cf [29] to [30] of Chapter III). Hence the elements of [171] can be written as

$$\sum_{\lambda \in \Lambda} (r_\lambda + i_\lambda 1) \square h_\lambda$$

or simply $\sum_{\lambda \in \Lambda} r_\lambda' \square h_\lambda$ (with $r_\lambda' \in R$) .

Dropping the prime for convenience, the above sum becomes

$$\sum_{\lambda \in \Lambda} r_\lambda \square h_\lambda \ .$$

Consequently, if both R and M are unitary then

$$\mathrm{Mod}((H)) = \left\{ \sum_{\lambda \varepsilon \Lambda} r_\lambda \,\square\, h_\lambda \;\middle|\; h_\lambda \varepsilon H,\ r_\lambda \varepsilon R,\ r_\lambda = 0 \text{ p.p. } \lambda \varepsilon \Lambda \right\}. \qquad [172]$$

Otherwise, we have

$$\mathrm{Mod}((H)) = [171]\ . \qquad [173]$$

(def) A subset B of a left R-module M is called *a set of generators* of M if $M = \mathrm{Mod}((B))$.

(def) A left R-module M is said to be *finitely generated* if M has a finite set of generators.

The concepts of "linear combinations" and "linear independence" play important roles in unitary R-modules. They are defined below.

(def) Let R be a unitary ring, and let M be a unitary left R-module. If $H = \{h_\lambda\}_{\lambda \varepsilon \Lambda}$ is any subset of M, then an *R-linear combination* of elements in H is defined as a sum

$$\sum_{\lambda \varepsilon \Lambda} r_\lambda \,\square\, h_\lambda \qquad [174]$$

with $h_\lambda \varepsilon H$, $r_\lambda \varepsilon R$ and $r_\lambda = 0$ p.p. $\lambda \varepsilon \Lambda$. The set of all R-linear combinations of elements in H is obviously identical to Mod((H)). This, as we noted before, is not true if not both

R and M are unitary. The set H is said to be R-linearly independent if, in the case of a finite set H,

$$\sum_{\lambda \varepsilon \Lambda} r_\lambda \square h_\lambda = 0, \; r_\lambda \varepsilon R \Longrightarrow r_\lambda = 0, \quad \text{for every } \lambda \varepsilon \Lambda. \qquad [17$$

If H is an infinite set, we say that H is R-linearly independent if every non-empty finite subset of H is R-linearly independent.

(def) Let M be a unitary left R-module, then a set of generators B (of M) is called a base (or basis, by some authors) of M if B is R-linearly independent. A base is said to be finite if it is finite as a set.

Remarks

1) The following is an equivalent definition of a base: a set of generators B (of M) is a base of M if, for every $x \varepsilon M$, the decomposition

$$x = \sum_{\lambda \varepsilon \Lambda} r_\lambda \square b_\lambda \quad \text{with} \quad r_\lambda \varepsilon R, \; b_\lambda \varepsilon B \qquad [17$$

is unique. The proof of the equivalence of the definitions is trivial for a finite base B. Let there be two such decompositio namely [176] and the following one:

$$x = \sum_{\lambda \varepsilon \Lambda} r_\lambda' \square b_\lambda \quad \text{with} \quad r_\lambda' \varepsilon R, \; b_\lambda \varepsilon B, \qquad [17$$

then, by [176] and [177],

$$\sum_{\lambda \in \Lambda} (r_\lambda - r_\lambda') \Box b = 0 . \quad [178]$$

[178] implies, by R-linear independence of B, $r_\lambda = r_\lambda'$, $\forall \lambda \in \Lambda$. We leave the case of an <u>infinite</u> base B to the reader (Hint: notice the difference in the definitions of <u>R-linear independence</u> for a <u>finite</u> set and an <u>infinite</u> set).

2) It is obvious, from the definition, that a subset B of M is a <u>base</u> of M iff B is <u>maximal</u> as an R-linearly independent subset of M (we leave the proof to our reader). ∎

<u>Proposition XV</u>

If a unitary left R-module M has a finite base then

R is a left Noether ring $\Longrightarrow$ M is Noetherian. [179]

and similarly,

R is a left Artin ring $\Longrightarrow$ M is Artinian. [180]

<u>Proof</u>

We shall prove only the Noetherian case since the Artinian statement can be proved in a similar manner.

Let $\{b_1, \ldots, b_n\}$ be a finite base for M, and consider the submodule

$$R \square b_i = \{ r \square b_i \mid r \in R \} \qquad [181]$$

for an arbitrary, fixed i. Clearly any ascending chain of submodules of $R \square b_i$ takes the form

$$R_1 \square b_i \subset R_2 \square b_i \subset \cdots \qquad [182]$$

where R_j are some subrings of R. It is quite easy to see that R_j are <u>left</u> ideals of R. For every $r \in R$ and $r_j \in R_j$, we have

$$(rr_j) \square b_i = r(r_j \square b_i) \in r(R_j \square b_i) \subset R_j \square b_i \qquad [183]$$

where we have used, in the last step, the fact that $R_j \square b_i$ is a <u>submodule</u> of $R \square b_i$ by assumption. [183] implies that

$$rr_j \in R_j \ , \qquad [184]$$

hence R_j is a <u>left</u> ideal of R. Therefore to each chain [182] there corresponds a chain (of left ideals of R):

$$R_1 \underset{\text{left}}{\subsetneq} R_2 \underset{\text{left}}{\subsetneq} \cdots . \qquad [185]$$

But R is left-Noetherian hence the chain [185] stops at a <u>finite</u> length. Consequently the chain [182] must also be finite. In other words, the submodule $R \square b_i$ satisfies the ACC, hence it is <u>Noetherian</u>.

Finally, since both R and M are unitary and since $\{b_1, \ldots, b_n\}$ is a base for M, we have

$$M = \sum_{i=1}^{n} R \,\square\, b_i \,. \qquad [186]$$

But each $R \,\square\, b_i$ is Noetherian as we have just shown, therefore, by Proposition XIII, M is Noetherian. ▮

The following property is one of the important differences between a Noetherian module and an Artinian module.

Proposition XVI

Let M be a left R-module, then

$$M \text{ is Noetherian} \Longleftrightarrow \text{every submodule of } M \text{ is finitely generated.} \qquad [187]$$

Proofs

Proof in the "$\Longrightarrow$" direction.

For any submodule M' of M, let us denote by Σ the collection of all finitely generated submodules of M'. Since M is Noetherian we see, by Proposition XII, that M' is also Noetherian. Thus M' satisfies the maximum condition. Therefore, Σ, as a non-empty collection (since at least $\{0\} \in \Sigma$) has a maximal member, to be denoted by M_o'. One can easily show that $M_o' = M'$. For any $x' \varepsilon M'$, the set $M_o' + \{x'\}$ is finitely generated since M_o' (as a member of Σ) is. In other words.

$$M_o' + \{x'\} \varepsilon \Sigma . \qquad [188]$$

But M_o' is maximal in Σ hence

$$M_o' + \{x'\} = M_o' \qquad [189]$$

i.e.

$$x' \varepsilon M_o' . \qquad [190]$$

Remember that x' is an arbitrary element of M'. Therefore, [190] implies $M_o' = M'$ as we claimed. Consequently, M' is finitely generated since M_o' is. ||

Proof in the "$\Longleftarrow$" direction.

Now we assume that every submodule of M is finitely generated. Consider any ascending chain, Σ, of submodules of M:

$$M_1 \subset M_2 \subset \cdots \quad . \qquad [191]$$

Then the set-theoretical union

$$\bigcup_{i=1}^{\infty} M_i \equiv M_U \qquad [192]$$

is obviously a <u>sumbodule</u> of M (though set-theoretical union does not lead to a submodule in general yet this is true for a <u>chain</u>). $M_{\cup}$ is , therefore, finitely generated by the assumption of [187]. Denote the finite set of generators by $X = \{x_1, \ldots, x_n\}$. We can write, by [192]

$$x_i \in M_{k_i} \qquad [193]$$

where k_i are some positive integers. In other words, corresponding to the set $\{ x_1, \ldots, x_n\}$ we have an index set

$$\{k_1, \ldots, k_n\} \qquad [194]$$

whose <u>maximal</u> element we denote by k, say. Then, by [191],

$$M_{k_i} \subset M_k$$

i.e. $x_i \in M_k$, for every $i \in I_+$

or $X \subset M_k$. [195]

But X generates $M_{\cup}$ which contains all M_i of Σ , therefore $M_i \subset M_k$ for every $i \in I_+$. Consequently the ascending chain Σ must stop at the finite length k. Thus M is Noetherian. ∎

§. 4.7. <u>Simplicity, semi-simplicity and composition series</u>

We have seen that the notions of "simplicity" and "semi-simplicity" have facilitated the study of group and ring structures. This is also the case with modules. In this section, R denotes a ring. When M denotes a left R-module, its module composition will be denoted by $\Omega_{\square}$ unless otherwise specified.

(def) A left R-module M is *simple* (i.e. *irreducible*) if M is non-trivial and if M has no non-zero proper submodule.

(def) A left R-module M is said to be *non-simple* (i.e. *reducible*) if M is not simple.

(def) A left R-module M is said to be *semi-simple* (i.e. *completely reducible*) if every submodule of M is a direct summand of M. (i.e. for every submodule M_1 of M there always exists a submodule M_2 of M such that $M = M_1 \oplus M_2$). Otherwise we say that M is *non-semi-simple*.

(def) A left R-module M is said to be *indecomposable* if M is non-trivial and is not expressible as a direct sum of non-zero submodules. Otherwise we say that M is *decomposable*.

Remarks

i) It is clear that, for a unitary M, "non-trivial" implies "non-zero".

ii) It follows from the definitions that (for modules):

$$\text{simplicity} \underset{\not\Leftarrow}{\Rightarrow} \text{semi-simplicity.} \qquad [196]$$

$$\text{simplicity} \underset{\not\Leftarrow}{\Rightarrow} \text{indecomposability.} \qquad [197]$$

$$\text{simplicity} \Leftrightarrow \text{semi-simplicity } \underline{\text{and}} \text{ indecomposability.} \qquad [198]$$

Example

The set I as an I-module is indecomposable and non-semi-simple. ▮

The concept of "torsion-free" of a module , similar to that of a ring, is defined here.

(def) An element x of a left R-module M is said to be *torsion-free* if

$$r \square x = 0, \text{ some } r \in R \Rightarrow r = 0 \,. \qquad [199]$$

The module M is said to be *torsion-free* if every non-zero element (of M) is torsion-free.

Proposition XVII

Let R be an integral domain, and let M be a unitary non-zero principal left R-module. Then M is indecomposable if it is torsion-free.

Proof

A unitary non-zero principal left R-module, M, has the form

$$M = R \square x \qquad [200$$

where x is a generator of M. First we want to show that M is R-isomorphic to the left R-module R (that a ring R can always be made into a left R-module was discussed in §.4.1. cf. [31] there). For clarity, let us denote the module composition in R (as a left R-module) by $\Omega_{\blacksquare}$, then it is understood

$$r' \blacksquare r = r'r \qquad (r, r' \varepsilon R) \qquad [201$$

where the LHS of [201] is a ring multiplication. The R-iso is established by the mapping

$$f : R \longrightarrow R \square x$$

with $$f : r \longmapsto r \square x , \qquad [202$$

where $r \varepsilon R$, and x is the generator of M (since M is principal). That [202] provides an R-iso can be easily proved. First, f is an R-hom since, for every r, r' ε R,

$$f(r' \blacksquare r) = f(r'r) = (r'r) \square x = r' \square (r \square x) = r' \square (fr) \qquad [203$$

and since $f(r' + r) = fr' + fr$. Next, the mapping [202] is clearly surjective. Finally, [202] is <u>injective</u> since if there are $r, r' \in R$ such that

$$r \square x = r' \square x$$

i.e. $$(r - r') \square x = 0 \qquad [204]$$

then it is necessary that

$$r = r'$$

in virtue of the torsion-free assumption of M (remember that $M = R \square x$).

The R-iso just established between M and R (as modules) reduces the proof of the theorem to that of <u>indecomposibility</u> of R (as a left R-module). This can be proved by method of contradition. Assume that R is decomposable, say,

$$R = R_1 \oplus R_2 \qquad \text{(R-module direct sum)} \qquad [205]$$

where R_1 and R_2 are non-zero left ideals (actually <u>ideals</u>) of R. For any non-zero $r_1 \in R_2$ and non-zero $r_2 \in R_2$, we have

$$r_1 r_2 = r_2 r_1 \in R_1 \cap R_2 = \{0\} \text{ (since } R_1 \text{ and } R_2 \text{ are ideals)} \qquad [206]$$

which contradicts the fact that R is an integral domain. Therefore R is indecomposable. ||

(def) A normal series of a left R-module M is defined as a finite chain of distinct submodules (of M),

$$M \equiv M_0 \underset{\text{mod}}{\supset} M_1 \underset{\text{mod}}{\supset} \cdots \underset{\text{mod}}{\supset} M_n \equiv \{0\} . \qquad [207]$$

(def) A normal series is a composition series if every quotient of the series, i.e. M_i/M_{i+1} of [207], is simple. These quotients are called composition quotients (i.e. composition factors).

(def) Two composition series, Σ and Σ', of a left R-module M are said to be isomorphic (i.e. equivalent by some authors) if they have the same length and if a 1-1 correspondence can be established be ween the terms of Σ and those of Σ' in such a way that their corresponding composition quotients are R-isomorphic. (cf. §.2.12)

The notion of "refinement" of a chain is also similar to that defined in Chapter II (on groups). Hence we shall not repeat it here.

Schreier Theorem takes the following form for modules (cf. Proposition XIX of Chapter II, on groups).

Proposition XVIII (Schreier's theorem)

Any two ascending (or two descending) finite chains of a left R-module possess R-isomorphic refinements.

Proof

The proof is exactly the same as in group theory. We prepare here a "dictionary" to translate the proof there to the case of a left R-module M (we urge the reader to copy this down on a piece of paper and use it to read the proof of Proposition XIX of Chapter II):

On groups (Chapter II)		For modules
group	$\longrightarrow$	a left R-module
N_i	$\longrightarrow$	M_i
G	$\longrightarrow$	M
1_G	$\longrightarrow$	0
$*$	$\longrightarrow$	$+$
normal series	$\longrightarrow$	ascending finite chains
$\prec$	$\longrightarrow$	$\underset{\text{mod}}{\subset}$
Zassenhaus lemma on groups	$\longrightarrow$	Zassenhaus lemma on modules (Proposition IX, §.4.4)
isomorphic	$\longrightarrow$	R-isomorphic. ∎

Again, as in group theory, Schreier theorem implies at once the Jordan-Hölder theorem on composition series. (cf. §.2.13)

Proposition XIX

Any two composition series of a left R-module are R-isomorphic.

Proof

Since a composition series is an ascending (or descending, if we reverse the writing) finite chain (by definition) Schreier theorem says that any two composition series have R-isomorphic refinements. However, a composition series has only trivial refinements, i.e. the series itself, due to the requirement of "simple" composition quotients. Thus any two composition series must be R-isomorphic. ||

The following proposition is a statement of the existence condition for a composition series:

Proposition XX

Let M be a left R-module, then

M has a composition series $\Longleftrightarrow$ M satisfies both the ACC and the DCC.

Proof

The truth in the "$\Longrightarrow$" direction is obvious.

For the "$\Longleftarrow$" direction, let us use the maximum condition since it is equivalent to the ACC.

If $M = \{0\}$ there is nothing to prove. If $M \neq \{0\}$, then by the maximum condition there is a maximal member M_1 in the collection of all proper submodules of M. Again, if $M_1 \neq \{0\}$, then there is a maximal member M_2 in the collection of all proper submodules of M_1. We carry out this repeatedly to get a descending chain of distinct submodules of M:

$$M \underset{\text{mod}}{\supset} M_1 \underset{\text{mod}}{\supset} M_2 \underset{\text{mod}}{\supset} \cdots \qquad [208]$$

with simple quotients (why? Hint: since no proper submodule of M_i can contain M_{i-1}). Besides, [208] must be of <u>finite</u> length by the DCC. Consequently we have obtained a composition series of M. ||

<u>Proposition XXI</u>

Every simple left R-module is principal.

<u>Proof</u>

Let M be a simple left R-module, then (by definition)

$$R \square M \neq \{0\} . \qquad [209]$$

Consider the set,

$$Y \equiv \{ y \mid y \in M, R \square y = \{0\} \} \qquad [210]$$

i.e.

$$R \square Y = \{0\} . \qquad [211]$$

Then [209] yields

$$R \square M \neq R \square Y$$

i.e. $$M \neq Y \, . \qquad [212]$$

It is clear that Y is a left R-module of M. Hence, by [212], Y is a *proper* left R-module of M. But M is *simple*, consequently

$$Y = \{0\} \, . \qquad [213]$$

In other words,

$$R \square x \neq \{0\} \, , \quad \text{for any non-zero } x \in M \, . \qquad [214]$$

Since $R \square x$ is a left R-module (of M) which is non-zero (by [214]) therefore

$$R \square x = M \, , \qquad [215]$$

in virtue of the *simplicity* of M. In other words, M is generated by a singleton x, hence M is principal. ||

Proposition XXII

Let M be a simple left R-module then, for any given $x \in M$, ann.x is modular and maximal as a submodule of the left R-module R.

Proof

R is considered here also as a *left R-module* by using the

<u>ring</u> multiplication as the <u>module</u> composition (as stated by [31]).

<u>Step 1</u>

By definition, for any given $x \in M$,

$$\text{ann.}x \equiv \{ a \mid a \in R,\ a \,\square\, x = 0 \} . \qquad [216]$$

Since (see §.4.2)

$$\text{ann.}x \subsetneq R , \qquad [217]$$

ann.x is a <u>submodule</u> of the left R-module R.

To show that ann.x is <u>maximal</u> let us assume there is a left ideal R' of R such that

$$R' \supsetneq \text{ann.}x . \qquad [218]$$

The mapping

$$f : R \longrightarrow M$$

with $\quad f : r \longmapsto r \,\square\, x , \quad$ for every $r \in R$, [219]

is obviously an R-epi. Since

$$fR' \underset{\text{mod}}{\subset} M$$

and since M is simple, it is necessary that

$$fR' = M \qquad (fR' \neq \{0\} \text{ by } [218])$$

i.e. $fR' = fR$.

Therefore, for every $r \in R$,

$$\exists\, r' \in R' : fr = fr' \qquad [22$$

i.e. $r \,\square\, x = r' \,\square\, x$

or $(r - r') \,\square\, x = 0$

or $(r - r') \in \text{ann.}x \in R'$.

Consequently $r \in R'$, i.e. $R \subset R'$, hence $R' = R$. In other word ann.x is maximal as a left R-module of R.

It is easy to see that ann.x is modular (we leave the proo to our reader. Or see the Problem Section). ||

Remark

It is clear that for a simple left R-module M,

$$\text{ann.}x = \text{ann.}M \text{ , for any non-zero } x \in M \text{ .} \qquad [22$$

Proposition XXIII

If M is a simple left R-module, then M is R-isomorphic to the quotient left R-module $R/\text{ann}.x$, where x is any non-zero element of M.

Proof

By [215] (of Proposition XXI),

$$M = R \,\square\, x \,, \quad \text{for any non-zero } x \in M \,. \qquad [222]$$

Since ann.x is a submodule of the left R-module R (see the statement below [217]) we can construct the quotient left R-module

$$R/\text{ann}.x \equiv \overline{R} \,. \qquad [223]$$

Introduce now the module composition

$$\Omega_{\blacksquare} : R \times \overline{R} \longrightarrow \overline{R}$$

with $\Omega_{\blacksquare} : (r', \overline{r}) \longmapsto \overline{r'r}$, for every $r, r' \in R$, [224]

where $\overline{r} \equiv r \bmod \text{ann}.x$. Then $\overline{R}$ is clearly a left R-module w.r.t. $\Omega_{\blacksquare}$. The mapping

$$f_* : \overline{R} \longrightarrow M$$

with $f_* : \overline{r} \longmapsto r \,\square\, x$, for every $r \in R$, [225] is an R-iso. f_* is obviously an R-hom since, for every $r_1, r_2 \varepsilon R$,

$$f_*(\overline{r}_1 + \overline{r}_2) = f_*(\overline{r_1 + r_2}) = (r_1 + r_2) \square x$$
$$= r_1 \square x + r_2 \square x = f_*(\overline{r}_1) + f_*(\overline{r}_2) \qquad [226$$

and
$$f_*(r_1 \cdot \overline{r}_2) = f_*(\overline{r_1 r_2}) = (r_1 r_2) \square x$$
$$= r_1 \square (r_2 \square x) = r_1 \square (f_* \overline{r}_2) . \qquad [227$$

The mapping f_* is clearly bijective : the surjectivity follows from the definition, and the injection follows from $\ker . f_* = \text{ann}.x = \overline{0}$.

||

Remark

We use the notation f_* to indicate the fact that f_* is induced from the f of [219] by "passing to the quotient".

Proposition XXIV

Every submodule of a semi-simple left R-module is semi-simple.

Proof

Let M' be a submodule of a semi-simple left R-module M, then the semi-simplicity of M implies that M' has a direct supplement, M", in M:

$$M = M' \oplus M'' . \qquad [228]$$

For any submodule W of M', by the semi-simplicity of M, W has a direct supplement W' in M,

$$M = W \oplus W' . \qquad [229]$$

clearly,

$$M' = M' \cap M = M' \cap (W \oplus W') = W \oplus (M' \cap W'). \qquad [230]$$

In other words, every submodule of M' has a direct supplement (in M'). Hence M' is semi-simple. ||

Proposition XXV

Let M be a left R-module. Then the following conditions are equivalent:

i) M is a finite sum of simple submodules (of M).

ii) M is a finite direct sum of simple submodules (of M).

iii) M is semi-simple and M has a composition series.

Proof

We shall carry out the proof according to the following scheme:

$$(i) \Longrightarrow (ii) \Longrightarrow (iii) \Longrightarrow (i) .$$

Proof of (i) $\Longrightarrow$ (ii).

By (i), we write **M** as a finite sum of simple submodules:

$$M = \sum_{i=1}^{n} M_i \ . \qquad [231$$

For any $i \neq j$ we have, by Proposition I,

$$M_i \cap M_j \subset_{\text{mod}} M_i \text{ (and } M_j \text{ too).} \qquad [232$$

But M_i is simple, $M_i \cap M_j$ must either be M_i or $\{0\}$. It cannot be M_i since $i \neq j$ (i.e. we exclude $M_i \cap M_i$ as well as $M_i \subset M_j$ in which case M_i should not appear in the sum [231]. Therefore

$$M_i \cap M_j = \{ 0 \} \ , \text{ for any } i \neq j \qquad [233$$

i.e. the sum [231] is direct. ||

Proof of (ii) $\Longrightarrow$ (iii).

By (ii) , we write M as a finite direct sum of simple submodules:

$$M = \bigoplus_{i=1}^{n} M_i \ . \qquad [234$$

Let M' be any submodule of M. Let us also assume that the

submodules, M_i's, in the direct sum [234] are labelled such that

$$M_i \not\subset M' , \quad \text{for every} \quad i > m . \tag{235}$$

In the extreme cases, i.e. $M' = M$ and $M' = \{0\}$, then $m = 0$ and $m = n$, respectively. Clearly,

$$\bigoplus_{i=1}^{m} M_i \underset{\text{mod}}{\subset} M' \tag{236}$$

hence

$$M = M' \oplus M'', \quad \text{where} \quad M'' \equiv \bigoplus_{i=m+1}^{n} M_i . \tag{237}$$

In other words M is semi-simple. ▌

It is not difficult to see that M has a composition series because

$$M \supset \bigoplus_{i=1}^{n-1} M_i \supset \cdots \supset M_2 \oplus M_1 \supset M_1 \supset \{0\} \tag{238}$$

is such a series. Each composition quotient

$$\bigoplus_{i=1}^{j} M_i \Big/ \bigoplus_{i=1}^{j-1} M_i \tag{239}$$

is simple since M_j is. ||

<u>Proof of (iii) $\Longrightarrow$ (i)</u>.

Let Γ be the collection of all submodules (of M) that are <u>not</u> expressible as finite sums of <u>simple</u> submodules of M. We want to show that Γ is empty if M is semi-simple and if M has a composition series.

Since M has a composition series, it follows, by Proposition XX, that M satisfies both chain condition. In particular, M satisfies the DCC which is equivalent to the <u>minimum condition</u>. Now, if Γ is <u>empty</u> then the proof is done. If Γ is non-empty then, by the minimum condition, Γ has a minimal element, say W. Clearly, $W \in \Gamma$ implies that W is non-simple and non-zero, i.e.

$$\exists W' \underset{\text{mod}}{\subset} M : \{0\} \subset W' \subset W . \qquad [240]$$

But M is semi-simple therefore, by the preceding proposition, W is also semi-simple, i.e.

$$\exists W'' \underset{\text{mod}}{\subset} M : W = W' \oplus W'' . \qquad [241]$$

However, W is a <u>minimal</u> member of Γ, hence $W', W'' \notin \Gamma$. In other words, both W' and W" are expressible as finite sums of some <u>simple</u> submodules (of M). It follows, in virtue of [241] that W is also expressible as a finite sum of simple submodules. That is

$$W \notin \Gamma \qquad [242]$$

which is absurd. Consequently $\Gamma = \emptyset$. So every submodule of M is expressible as a finite sum of simple submodules (of M). In particular, M is express ble as such a sum. ||

§. 4.8. Free modules

R denotes a unitary ring in this section.

The discussion, in this section, on free R-modules is essentially similar to that on free groups or free monoids. Their common abstractions and features allow one to deal with several different algebraic structures at the same time in this context. However, a module that is "free" enjoys many nice properties and, therefore, plays an important role envied by both a free group or a free monoid.

(def) Let M be a left R-module, and let H be an arbitrary set together with a mapping $\alpha \in \mathrm{Map}(H, M)$, then the pair $\{M, \alpha\}$ is called a free left R-module on H if, for any left R-module W and a mapping $\beta \in \mathrm{Map}(H, W)$,

$$\exists_{1!}\ \mu \in \mathrm{Hom}_R(M, W): \mu \circ \alpha = \beta \qquad [243]$$

i.e. the following diagram is <u>commutative</u>:

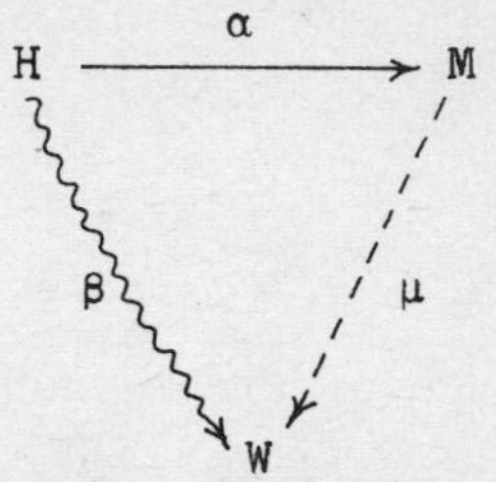

[244]

<u>Remark</u>

When $\{M, \alpha\}$ is a <u>free</u> left R-module on a set H we may simply say "the left R-module M is <u>free</u> on H" or "H generates the free left R-module M". In general, when there is no necessity of mentioning the mapping α we shall write M in place of $\{M, \alpha\}$.

Similar to the case of a monoid or a group, the following property can be easily shown (we leave this to the reader).

<u>Proposition XXVI</u>

Let $\{M_1, \alpha_1\}$ and $\{M_2, \alpha_2\}$ be two <u>free left R-modules</u> on the same set H, then

$$\underset{1!}{\exists}\ \mu \in \mathrm{Iso}_R(M_1, M_2) : \mu \circ \alpha_1 = \alpha_2 \qquad [245]$$

i.e. the following diagram is <u>commutative</u>:

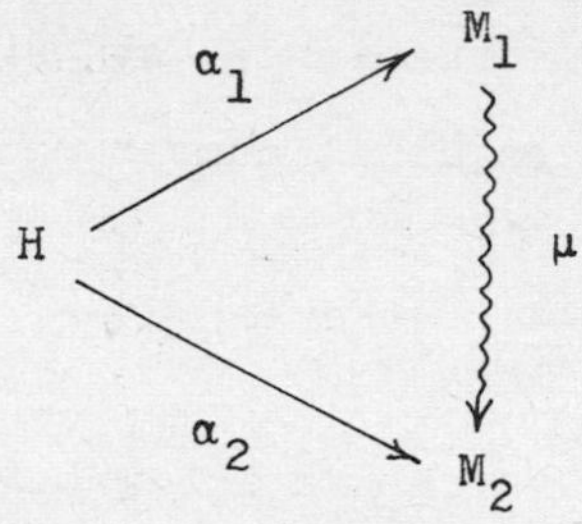

[246]

:onstruction of a free left R-module.

The construction of a free module on a set, to be discussed ·elow, is very similar to the case of a free monoid or a free group.

Let $H = \{h_\lambda\}_{\lambda \varepsilon \Lambda}$ be an arbitrary set (we disregard any possible xtra structures on H) and let the unitary ring R be written dditively and multiplicatively. We have seen that ordered sets ere introduced in the case of a monoid or a group. Each ordered et contains an arbitrary but finite number of elements from H, .g.

$$\{h_1, \ldots, h_m\}, \quad h_\lambda \varepsilon H \qquad [247]$$

ıere $h_i = h_j$ is not excluded. However, to endow H with a left -module structure we shall use, instead of [247], the following :dered set:

$$\{r_1 h_1, \ldots, r_m h_m\} \qquad [248]$$

ıere $r_i \varepsilon R$, and $r_i = r_j$ is not excluded. The ordered sets

[248] can be replaced, for convenience, by a formal sum (possibl
infinite, depending on the cardinal number of the set H):

$$\sum_{\lambda \varepsilon \Lambda} r_\lambda h_\lambda \ . \qquad [24\text{\textbackslash}$$

We must emphasize that the attachment of r_λ to h_λ to form r_λ
has nothing to do with any multiplication commitment between r_λ and h_λ. They are simply forced together to stay in the same "pigeon-hole" λ. Since H is not necessarily a finite set we impose here the usual condition for a sum:

$$r_\lambda = 0 \ , \quad \text{p.p. } \lambda \varepsilon \Lambda \ . \qquad [25$$

Now we introduce the following two compositions:

addition, $$\sum_\lambda r_\lambda h_\lambda + \sum_\lambda r_\lambda' h_\lambda = \sum_\lambda (r_\lambda + r_\lambda')h_\lambda \qquad [25$$

and module composition, $\Omega_\square$,

$$r \,\square \sum_\lambda r_\lambda h_\lambda = \sum_\lambda (r r_\lambda) h_\lambda \ , \qquad [2$$

for any $r, r_\lambda, r_\lambda' \varepsilon R$ and $h_\lambda \varepsilon H$. Then the collection, M, of all the formal sums defined by [249] and [250], is a left R-module w.r.t. the addition and module composition defined abo
To show this, let us go through the axioms of a left R-module:

) $r \,\square\, (\sum_{\lambda} r_{\lambda} h_{\lambda} + \sum_{\lambda} r_{\lambda}' h_{\lambda}) = r \,\square\, \sum_{\lambda}(r_{\lambda} + r_{\lambda}')h_{\lambda} = \sum_{\lambda}(r(r_{\lambda} + r_{\lambda}'))h_{\lambda}$

$$= \sum_{\lambda}(rr_{\lambda} + rr_{\lambda}')h_{\lambda} = r \,\square\, \sum_{\lambda} r_{\lambda} h_{\lambda} + r \,\square\, \sum_{\lambda} r_{\lambda}' h_{\lambda} .$$

) $(r + r') \,\square\, \sum_{\lambda} r_{\lambda} h_{\lambda} = \sum_{\lambda}((r + r')r_{\lambda})h_{\lambda} = \sum_{\lambda}(rr_{\lambda} + r'r_{\lambda})h_{\lambda}$

$$= \sum_{\lambda}(rr_{\lambda})h_{\lambda} + \sum_{\lambda}(r'r_{\lambda})h_{\lambda} = r \,\square\, \sum_{\lambda} r_{\lambda} h_{\lambda} + r' \,\square\, \sum_{\lambda} r_{\lambda} h_{\lambda} .$$

i) $(rr') \,\square\, \sum_{\lambda} r_{\lambda} h_{\lambda} = \sum_{\lambda}((rr')r_{\lambda})h_{\lambda} = \sum_{\lambda}(r(r'r_{\lambda}))h_{\lambda}$

$$= r \,\square\, \sum_{\lambda}(r'r_{\lambda})h_{\lambda} = r \,\square\, (r' \,\square\, \sum_{\lambda} r_{\lambda} h_{\lambda}).$$

v) $1 \,\square\, \sum_{\lambda} r_{\lambda} h_{\lambda} = \sum_{\lambda}(1 \,\square\, r_{\lambda})h_{\lambda} = \sum_{\lambda} r_{\lambda} h_{\lambda}$. **||**

Next, we want to show that M is a <u>free</u> left R-module generated H. Consider the mapping α , of H into M, defined by:

$$\alpha : h_{\lambda} \longmapsto \sum_{\sigma \in \Lambda} \delta_{\lambda\sigma} h_{\sigma} \qquad [253]$$

ere $\delta_{\lambda\sigma}$ is defined by

$$\delta_{\lambda\sigma} = 0 \text{ , if } \lambda \neq \sigma \quad \text{and} \quad \delta_{\lambda\sigma} = 1 \text{ , if } \lambda = \sigma \qquad [254]$$

ere 0 and 1 are the zero and the multiplicative unit of the itary ring R. Then any element of M can be written <u>uniquely</u> as

$$\sum_{\lambda} (r_\lambda \square \sum_{\lambda} \delta_{\lambda\sigma} \; h_\sigma \;) \; .$$ [255

This implies, by definition, that M has a base αH whose element
are

$$\sum_{\sigma} \delta_{\lambda\sigma} h_\sigma \, , \; \lambda \in \Lambda \; .$$ [256

The proof that M is a free R-module generated by H proceeds i
essentially the same way as in the case of a free monoid; we wan
to show that, for any given left R-module W and $\beta \in \text{Map}(H, W)$,

$$\underset{1!}{\exists} \; \mu \in \text{Hom}_R(M, W) \; : \; \mu \circ \alpha = \beta \; .$$ [25

That is, the following diagram is commutative:

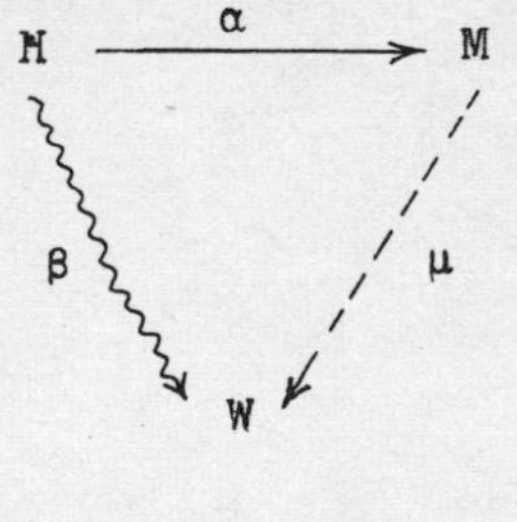

[25

The required R-hom μ is given by

$$\mu \; : \; \sum_{\lambda \in \Lambda} r_\lambda h_\lambda \longmapsto \sum_{\lambda \in \Lambda} r_\lambda (\beta h_\lambda) \, ,$$ [25

for every $r_\lambda \in R$ and $h_\lambda \in H$. Clearly

$$\mu\Big|_{\alpha H} = \beta \quad , \qquad [260]$$

hence

$$\mu \circ \alpha = \beta \quad .$$

||

Proposition XXVII

Let M be a left R-module, and let H be an arbitrary set with a mapping $\alpha \in \mathrm{Map}(H, M)$, then

$$\{M, \alpha\} \text{ is a } \underline{\text{free}} \text{ left R-module generated by H} \Longleftrightarrow$$
$$\alpha H \text{ is a } \underline{\text{base}} \text{ of M.} \qquad [261]$$

Proof

The proof in the ($\Longrightarrow$) direction is given by [255] (see the statement following [255]). We leave the proof in the direction of ($\Longleftarrow$) to the reader (or see the Problem Section). ||

Proposition XXVIII

Every left R-module is isomorphic to a quotient module of a free R-module.

Proof

Consider a left R-module M. Let H be a subset (of M) that generates M. H always exists since at least we can take $H = M$.

Let $\{F, \alpha\}$ be the free left R-module generated by H, then there is a unique $\mu \in \mathrm{Hom}_R(F, M)$ such that the following diagram is commutative:

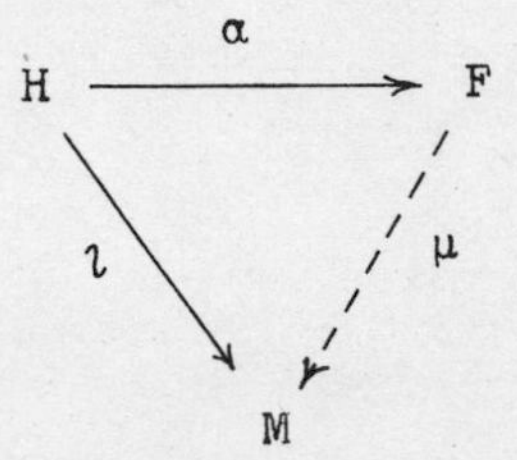

[262

where ι is the inclusion mapping. Clearly [262] requires that $\mu F \supset \iota H = H$. Since H generates M, i.e. M is the smallest left R-module containing H, and also since μ is an R-hom, this impli $\mu F \supset M$. In other words,

$$\mu F = M .$$

Hence $$M \longleftrightarrow F \bmod \ker.\mu \quad .$$ ∎ [26

§. 4.9. <u>Tensor product</u>

R denotes a *unitary* ring in this section.

Before we give the definition of a "tensor product" let us introduce the notion of a balanced map.

(def) Let M and N be, respectively, right and left R-modules .

Their module compositions will be denoted by the same symbol $\Omega_{\square}$, for economy. Let T be an abelian group written additively (M, N and T all share the same additive sign for economy of notation) then a mapping $\beta \in \mathrm{Map}(M \times N, T)$ is called a balanced mapping, from $M \times N$ to T, if the following three conditions are satisfied for every $x, x' \in M$, $y, y' \in N$ and $r \in R$:

i) $\beta(x, y + y') = \beta(x, y) + \beta(x, y')$ [264]

ii) $\beta(x + x', y) = \beta(x, y) + \beta(x', y)$ [265]

iii) $\beta(x \square r, y) = \beta(x, r \square y)$ [266]

(def) Tensor product of R-modules.

Let M, N and T be defined as in the above definition, and let β be a balanced mapping from $M \times N$ to T. Then the pair {T, β} is called a tensor product of M and N if, for every abelian group G and every balanced mapping α from $M \times N$ to G,

$$\underset{1!}{\exists}\, \mu \in \mathrm{Hom}(T, G) : \mu \circ \beta = \alpha .$$ [267]

In other words the following diagram is commutative:

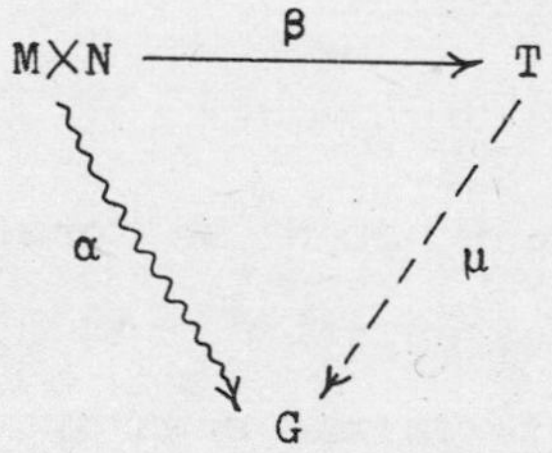

[268]

For simplicity, we often call T instead of the pair $\{T, \beta\}$ the tensor product of M and N.

Proposition XXIX

Let $\{T_1, \beta_1\}$ and $\{T_2, \beta_2\}$ be two tensor products of M and M'. Then

$$\exists_{1!} \nu \in \text{Iso}(T_1, T_2) : \nu \circ \beta_1 = \beta_2 .$$ [26

Proof

The proof is similar to the case of a free monoid or a fre group (cf. Proposition IV of Chapter I or Proposition XIII of Chapter II). ||

Remark

The above theorem points out that a tensor product is uniq up to an isomorphism (as an abelian group!). Therefore any particular construction of a tensor product is unique in this sense.

Construction of a tensor product $M \otimes N$

Step 1

From the cartesian product $M \times N$ we form the free I-modu $I(M, N)$, generated by all ordered pairs from $M \times N$. Therefore, $I(M, N)$ consists of all finite formal sums of the form

$$\sum_{\lambda \in \Lambda} i_\lambda(x_\lambda, y_\lambda) \quad [270]$$

with $x_\lambda \in M$, $y_\lambda \in N$, $i_\lambda \in I$ and $i_\lambda = 0$ p.p. $\lambda \in \Lambda$.

Step 2

Construct now $P(M, N)$, the sub I-module (of $I(M, N)$) generated by the set of all elements of the forms:

i) $(x, y + y') - (x, y) - (x, y')$ [271]

ii) $(x + x', y) - (x, y) - (x', y)$ [272]

iii) $(x \square r, y) - (x, r \square y)$ [273]

where $x, x' \in M$; $y, y' \in N$ and $r \in R$. Then we can show that the quotient I-module

$$I(M, N)/P(M, N) \quad [274]$$

is the required tensor product of the R-modules M and N. For convenience, we introduce the following notations:

$$M \otimes_R N = I(M, N)/P(M, N) \quad [275]$$

and $x \otimes y = (x, y) \bmod P(M, N)$, $x \in M$ and $y \in N$. [276]

Step 3

To show that $M \otimes_R N$ thus constructed is a tensor product let us introduce the "canonical" mapping

$$\beta : M \times N \longrightarrow M \otimes_R N \qquad [277]$$

with $\beta : (x, y) \longmapsto \beta(x, y) \equiv x \otimes y = (x, y) \bmod P$, [278]

where $x \in M$, $y \in N$ and P is the abbreviation of P(M, N). We have, for every $x, x' \in M$ and $y, y' \in N$,

$$\beta(x + x', y) = (x + x', y) \bmod P$$

$$\beta(x, y) = (x, y) \bmod P$$

and $$\beta(x', y) = (x', y) \bmod P .$$

It follows that

$$\beta(x + x', y) - \beta(x, y) - \beta(x', y) = \{(x + x', y) - (x, y) - (x', y)\} \bmod$$

But the curly bracket { } on the RHS of the above identity is an element of P, hence

$$\beta(x + x', y) - \beta(x, y) - \beta(x', y) = 0_{\otimes} . \qquad [279$$

where $0_{\otimes}$ denotes the zero of $M \otimes N$ under the abelian group structure. Similarly, by [271] and [272], we get

$$\beta(x, y + y') - \beta(x, y) - \beta(x, y') = 0_{\otimes} \qquad [28$$

and $\beta(x \square r, y) - \beta(x, r \square y) = 0_{\otimes}$. [281]

[279], [280] and [281] can be written as

$$\beta(x + x', y) = \beta(x, y) + \beta(x', y) \quad [282]$$

$$\beta(x, y + y') = \beta(x, y) + \beta(x, y') \quad [283]$$

$$\beta(x \square r, y) = \beta(x, r \square y) . \quad [284]$$

Therefore, by definition, β is a balanced mapping. By [278] we can write [282], [283] and [284] as

$$(x + x') \otimes y = x \otimes y + x' \otimes y \quad [285]$$

$$x \otimes (y + y') = x \otimes y + x \otimes y' \quad [286]$$

$$(x \square r) \otimes y = x \otimes (r \square y) . \quad [287]$$

To show that $I(M, N)/P(M, N)$ is a tensor product it remains only to verify [267]. Before we do this let us give some details about the left I-module structures in the next step.

Step 4

The left I-module structure is introduced to M in the obvious way. The module composition is given by

$$\Omega : I \times M \longrightarrow M \quad [288]$$

with $$\Omega : (t, x) \longmapsto tx \equiv \underbrace{x + \cdots + x}_{t \text{ copies}} \qquad [289]$$

for every $t \varepsilon I$ and $x \varepsilon M$. Similarly N is made a left I-module in the same way. From [282], [283] and [289] we have

$$\beta(tx, y) = t\beta(x, y) = \beta(x, ty) \qquad [290]$$

i.e. $$(tx) \otimes y = t(x \otimes y) = x \otimes (ty) \ . \qquad [291]$$

Now $M \otimes_R N$ can be made into a <u>left I-module</u> in the same way, with

$$I \times M \otimes_R N \longrightarrow M \otimes_R N \qquad [292]$$

defined by $$(t, \beta(x, y)) \longmapsto t\beta(x, y) \ , \qquad [293]$$

for every $t \varepsilon I$, $x \varepsilon M$ and $y \varepsilon N$. It is understood that

$$t\beta(x, y) = \underbrace{\beta(x, y) + \cdots + \beta(x, y)}_{t \text{ copies}} \ . \qquad [294]$$

Now we are in a position to look into $M \otimes_R N$. By definitio every element of $M \otimes_R N$ has the form

$$\sum_{\lambda \varepsilon \Lambda} i_\lambda \beta(x_\lambda, y_\lambda) \ , \quad i_\lambda \in I, \quad x_\lambda \in M, \quad y_\lambda \in N \qquad [295]$$

which, by [290], is equal to

$$\sum_{\lambda \varepsilon \Lambda} \beta(i_\lambda x_\lambda, y_\lambda) \quad . \tag{296}$$

This establishes that $M \otimes_R N$ is the abelian group <u>generated</u> by all the elements $\beta(x, y)$, $x \varepsilon M$, $y \varepsilon N$. ∎

<u>Step 5</u>

Finally, it remains to show that, for any abelian group G and any balanced mapping α from $M \times N$ to G, there exists a <u>unique</u> $\mu \varepsilon \mathrm{Hom}(M \otimes_R N, G)$ such that the following diagram is commutative:

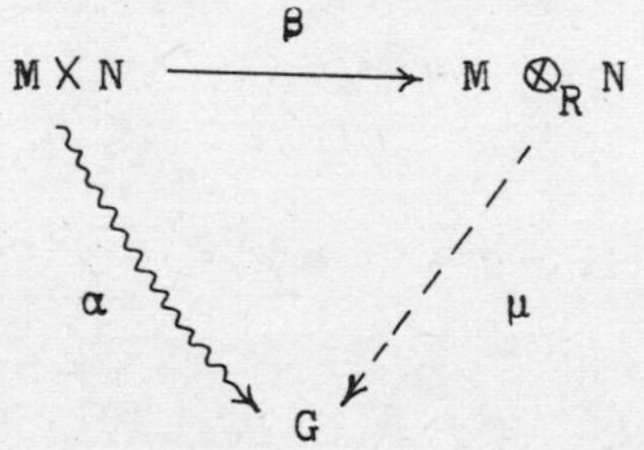

[297]

This, in fact, is quite obvious. Since $M \times N$ is a <u>base</u> of $I(M \times N)$ and since each element of $M \otimes_R N$ has the form

$$\sum_{\lambda} x_\lambda \otimes y_\lambda \tag{298}$$

the hom μ is <u>uniquely</u> determined by α through

$$\mu : x \otimes y \longmapsto \alpha(x, y), \qquad [29$$

for every $x \in M$ and $y \in N$. It is trivial to see that

$$\mu \circ \beta = \alpha \quad . \blacksquare \qquad [30$$

The special case where R is a <u>commutative</u> ring is of great importance; it permits us to impose a <u>left R-module</u> structure on $M \otimes_R N$ (remember that $M \otimes_R N$ has only a left I-module structu for an arbitrary R). The following simple analysis indicates wh this is so.

Let us denote by $\Omega_{\blacksquare}$ the composition

$$\Omega_{\blacksquare} : R \times M \otimes_R N \longrightarrow M \otimes_R N \qquad [30$$

with $\Omega_{\blacksquare} : (r, x \otimes y) \longmapsto r \blacksquare (x \otimes y) = x \otimes (r \square y) = (x \square r) \otimes y$

for every $r \in R$, $x \in M$ and $y \in N$. For [301] to be a <u>module</u> composition it is necessary that the following condition is satisfied:

$$(r'r) \blacksquare (x \otimes y) = r' \blacksquare (r \blacksquare (x \otimes y)) , \qquad [3$$

for every $r, r' \in R$, $x \in M$ and $y \in N$. First, by [301],

$$\text{LHS of [302]} = x \otimes ((r'r) \square y) = x \otimes (r' \square (r \square y)). \qquad [3$$

Next,

$$\text{RHS of } [302] = r' \blacksquare (x \otimes (r \square y)) = x \otimes (r' \square (r \square y)) \qquad [304]$$

Hence, in view of [303] and [304], it appears that [302] is satisfied whether R is commutative or not. But this is not so since we also have, by the last member of [301],

$$(r'r) \blacksquare (x \otimes y) = (x \square (r'r)) \otimes y \qquad [305]$$

and $$r' \blacksquare (r \blacksquare (x \otimes y)) = r' \blacksquare ((x \square r) \otimes y) = ((x \square r) \square r') \otimes y$$
$$= (x \square (rr')) \otimes y . \qquad [306]$$

clearly [305] is not equal to [306] in general if R is <u>not commutative</u>. When R is commutative then [301] is indeed a module composition which makes $M \otimes_R N$ a left R-module.

(def) Tensor product of R-homomorphisms

Let M and M' be right R-modules, N and N' be left R-modules. For any $f \in \mathrm{Hom}_R(M, M')$ and $g \in \mathrm{Hom}_R(N, N')$, the mapping $f \otimes g$ defined by

$$f \otimes g : M \otimes_R N \longrightarrow M' \otimes_R N',$$

with $$f \otimes g : x \otimes y \longmapsto (fx) \otimes (gy) , \; x \in M, \; y \in N, \qquad [307]$$

is called the *tensor product* of f and g. $f \otimes g$ is clearly an I-hom. In the case that R is commutative then $f \otimes g$ is also an

R-hom. This is analogous to the situation of $M \otimes_R N$. When the ground ring R is commutative, $M \otimes_R N$ has also a left R-module structure. A tensor product has the distributivity property in the following sense: let M be a right R-module decomposable into a direct sum of submodules:

$$M = \bigoplus_{i=1}^{n} M_i \, , \qquad [30$$

then for any left R-module N,

$$M \otimes_R N \longleftrightarrow \bigoplus_{i=1}^{n} (M_i \otimes N) \; . \qquad [30$$

We leave the proof to the reader (or see the Problem Section where the details are given).

§. 4.10. Exact sequences, extensions, projective and injective modules

R denotes a unitary ring in this section. All R-modules considered here are unitary left R-modules and will be referred to simply as "R-modules".

(def) Exact sequence

Let M_1, M_2, ..., M_n be R-modules with R-homomorphisms

$$f_i \in \mathrm{Hom}_R(M_i, M_{i+1}), \quad i=1, \ldots, n-1, \qquad [310]$$

then the sequence

$$M_1 \xrightarrow{f_1} M_2 \xrightarrow{f_2} \cdots \xrightarrow{f_{n-1}} M_n \qquad [311]$$

is said to be exact if

$$\mathrm{im}.f_i = \ker.f_{i+1}, \quad \text{for } i = 1, \ldots, n-2. \qquad [312]$$

(def) An exact sequence of R-modules of the form

$$0 \longrightarrow M \xrightarrow{\alpha} L \xrightarrow{\mu} N \longrightarrow 0 \qquad [313]$$

is called an extension of M by N. Frequently, we also say that L is an extension of M by N. An extension is also called a "short" exact sequence, by many authors.

Proposition XXX

a) The sequence (of R-modules)

$$0 \xrightarrow{\alpha} M \xrightarrow{\beta} M'$$

is exact $\Longleftrightarrow$ $\beta \in \mathrm{Mon}_R(M, M')$. [314]

b) The sequence

$$M \xrightarrow{\gamma} M' \xrightarrow{\sigma} 0$$

is exact $\Longleftrightarrow \gamma \in Epi_R(M, M')$. [315

Proof

The proofs are the same as in group theory therefore we shall not repeat here. ||

Remarks

i) By the above proposition, we have the equivalent definiti of an extension: an exact sequence of R-modules

$$M \xrightarrow{\alpha} L \xrightarrow{\mu} N$$ [316

is an extension of M by N if $\alpha \in Mon_R(M, L)$ and $\mu \in Epi_R(L,$

ii) A diagram is said to be row-exact if each row (i.e. a horizontal file) of the diagram forms an exact sequence, and column-exact if each column (i.e. a vertical file) forms an exac sequence.

(def) Two extensions (of R-modules)

$$0 \longrightarrow M \xrightarrow{\alpha} L \xrightarrow{\mu} N \longrightarrow 0$$ [31

and $$0 \longrightarrow M \xrightarrow{\alpha'} L' \xrightarrow{\mu'} N \longrightarrow 0$$ [31

are said to be equivalent (i.e. isomorphic) if

$$\exists f \in Hom_R(L, L') : \quad f \circ \alpha = \alpha' \quad \text{and} \quad \mu' \circ f = \mu \quad .$$

In other words, the following diagram is commutative:

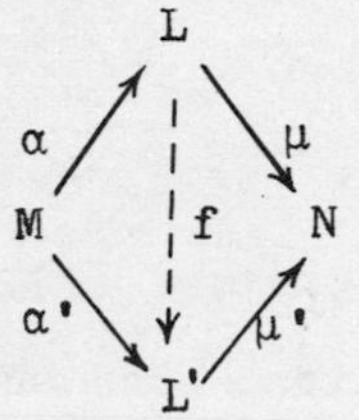

[319]

It is easy to see that f is actually an R-iso.

(notation) Ext (M, N) ≡ the set of all equivalent classes of extensions of M by N.

(def) An extension (of R-modules) L, of M by N,

$$0 \longrightarrow M \xrightarrow{\alpha} L \xrightarrow{\mu} N \longrightarrow 0 \qquad [320]$$

is said to be split if ker.μ is a direct summand of M.

<u>Proposition XXXI</u>

An extension (of R-modules)

$$0 \longrightarrow M \xrightarrow{\alpha} L \xrightarrow{\mu} N \longrightarrow 0$$

is <u>split</u> $\Longleftrightarrow$

$$\exists \beta \in Epi_R(L, M) : \beta \circ \alpha = \hat{1}_M \qquad [321]$$

and $\exists \nu \in \mathrm{Mon}_R(N, L) : \nu \circ \mu = \hat{1}_L$ [322]

where $\hat{1}_M$ and $\hat{1}_N$ are idenity R-endomorphisms on M and N.

Proof

The proof is patterned after the corresponding proof in group theory (Proposition XXII, Chapter II). We shall give only the proof in the " $\Longrightarrow$ " direction. The sufficiency proof is left to the reader (or see the Problem Section).

Proof of [321]: By the definition of "split", we have an R-module L' such that

$$L = L' \oplus \ker.\mu \; . \qquad [323]$$

Hence any element $x \in L$ can be decomposed into

$$x = x' + x'' \quad \text{with} \quad x' \in L' \quad \text{and} \quad x'' \in \ker.\mu \; . \qquad [324]$$

The sought-after β of [321] is given by

$$\beta : x \longmapsto \alpha^{-1}(x'') \; . \qquad [325]$$

We leave the reader to check that β satisfies [321]. ▮

Proof of [322]: The required ν of [322] is simply

$$\nu \equiv \left(\mu|_{L'}\right)^{-1} \; . \qquad [326]$$

We leave the verification of this claim to the reader. (If the reader does not succeed then he is advised to go through "step 2" of Proposition XXII, Chapter II; all one has to do is to realize that the <u>multiplication</u> there corresponds to the <u>addition</u> here, etc.) ||

The following is a useful property of "exactness".

<u>Proposition XXXII</u> (the "5-lemma")

Let the following commutative diagram of R-modules be <u>row-exact</u>:

$$\begin{array}{ccccccccc} M_2 & \xrightarrow{\alpha_2} & M_1 & \xrightarrow{\alpha_1} & M_o & \xrightarrow{\alpha_o} & M_{-1} & \xrightarrow{\alpha_{-1}} & M_{-2} \\ \downarrow{\scriptstyle\gamma_2} & & \downarrow{\scriptstyle\gamma_1} & & \downarrow{\scriptstyle\gamma_o} & & \downarrow{\scriptstyle\gamma_{-1}} & & \downarrow{\scriptstyle\gamma_{-2}} \\ N_2 & \xrightarrow[\beta_2]{} & N_1 & \xrightarrow[\beta_1]{} & N_o & \xrightarrow[\beta_o]{} & N_{-1} & \xrightarrow[\beta_{-1}]{} & N_{-2} \end{array} \qquad [327]$$

then

i) $\text{coker}.\gamma_2 = \text{ker}.\gamma_1 = \text{Ker}.\gamma_{-1} = \{0\} \Longrightarrow \text{ker}.\gamma_o = \{0\}$, [328]

ii) $\text{coker}.\gamma_{-2} = \text{coker}.\gamma_1 = \text{coker}.\gamma_{-1} = \{0\} \Longrightarrow \text{coker}.\gamma_o = \{0\}$. [329]

<u>Proof</u>

<u>Proof of (i)</u>

<u>Step 1</u>

Chase the diagram along the square formed by $M_o\ M_{-1}\ N_{-1}\ N_o$. We have,

$$\gamma_{-1}\alpha_o = \beta_o\gamma_o .$$

i.e. $\gamma_{-1}(\alpha_o x_o) = \beta_o(\gamma_o x_o) = 0$, for every $x_o \varepsilon \ker.\gamma_o$

or $\alpha_o x_o = 0$ (since $\ker.\gamma_{-1} = \{0\}$)

i.e. $x_o \varepsilon \ker.\alpha_o$.

Hence $\ker.\gamma_o \subset \ker.\alpha_o$

$= \text{im}.\alpha_1$ ([327] has exact rows)

i.e. $\exists x_1 \varepsilon M_1 : \alpha_1 x_1 = x_o$. [330]

<u>Step 2</u>

Chase the diagram along $M_1M_oN_oN_1$. Then

$$\beta_1\gamma_1 = \gamma_o \alpha_1$$

i.e. $(\beta_1\gamma_1)x_1 = (\gamma_o\alpha_1)x_1 = \gamma_o x_o = 0$.

Hence $\gamma_1 x_1 \varepsilon \ker.\beta_1 = \text{im}.\beta_2$

i.e. $\exists y_2 \varepsilon N_2 : \beta_2 y_2 = \gamma_1 x_1$. [331]

But $\text{coker}.\gamma_2 = \{0\}$ implies that $y_2 = \gamma_2 x_2$, for some $x_2 \varepsilon M$
Therefore [331] becomes

$$(\beta_2\gamma_2)x_2 = \gamma_1 x_1 .$$ [332

Finally, let us chase the diagram along $M_2M_1N_1N_2$. Then

$$\gamma_1\alpha_2 = \beta_2\gamma_2$$

By [332], we have

$$\gamma_1 x_1 = (\gamma_1\alpha_2)x_2$$

i.e. $x_1 = \alpha_2 x_2$. (since $\ker.\gamma_1 = \{0\}$)

or $\alpha_1 x_1 = \alpha_1\alpha_2 x_2 = 0$ ([327] has exact rows)

which implies, by [330], that

$$\ker.\gamma_0 = \{0\} .$$ I

We leave the proof of (ii) to the reader. II

Remark

[328] and [329] can be stated in a verbal way by means of the facts:

$$\text{coker.}f = \{0\} \Longrightarrow f \text{ is an R-epi. ,} \qquad [333]$$

$$\ker.f = \{0\} \Longrightarrow f \text{ is an R-mono.} \qquad [334]$$

Proposition XXXIII (the "short 5-lemma")

Let the following commutative diagram (of R-modules) be

<u>row-exact</u>:

$$\begin{array}{ccccccccc} 0 & \longrightarrow & M_1 & \xrightarrow{\alpha_1} & M_o & \xrightarrow{\alpha_o} & M_{-1} & \longrightarrow & 0 \\ & & \downarrow{\scriptstyle \gamma_1} & & \downarrow{\scriptstyle \gamma_o} & & \downarrow{\scriptstyle \gamma_{-1}} & & \\ 0 & \longrightarrow & N_1 & \xrightarrow[\beta_1]{} & N_o & \xrightarrow[\beta_o]{} & N_{-1} & \longrightarrow & 0 \end{array} \qquad [335]$$

then γ_o is <u>R-iso</u> (or respective <u>R-mono</u> and <u>R-epi</u>) if both γ_1 and γ_{-1} are.

<u>Proof</u>

By [328], it is clear that γ_o is <u>R-mono</u> if both γ_1 and γ_{-1} are. By [329], γ_o is R-epi if both γ_1 and γ_{-1} are. Combining these two facts we dispose of the <u>R-iso</u> case. ‖

(def) An R-module P is said to be *projective* if, for every row-exact diagram

$$\begin{array}{ccccc} M & \xrightarrow{\mu} & N & \longrightarrow & 0 \\ & & \uparrow{\scriptstyle \alpha} & & \\ & & P & & \end{array} \qquad [336]$$

there exists $\beta \in \mathrm{Hom}_R(P, M)$ such that the following diagram is commutative:

$$\begin{array}{ccccc} M & \xrightarrow{\mu} & N & \longrightarrow & 0 \\ & \nwarrow{\scriptstyle \beta} & \uparrow{\scriptstyle \alpha} & & \\ & & P & & \end{array} \qquad [337]$$

i.e. $\mu \circ \beta = \alpha$. [338]

Proposition XXXIV

Every free R-module is projective (the converse is not true).

Proof

Let P be a free module then P has a base. Denote by $\{y_\lambda\}_{\lambda \varepsilon \Lambda}$ a base of P. Consider the diagram [337]. We want to find β . Define the sets

$$X_\lambda \equiv \{x_\lambda \mid x_\lambda \varepsilon M , \mu x_\lambda = \alpha y_\lambda\} , \lambda \varepsilon \Lambda , \quad [339]$$

then every X_λ is non-empty since μ is an R-epi by [337]. For every $p \varepsilon P$, we can write (in terms of the base):

$$p = \sum_{\lambda \varepsilon \Lambda} a_\lambda \square y_\lambda , \; a_\lambda \varepsilon R \text{ and } a_\lambda = 0 \text{ p.p. } \lambda \varepsilon \Lambda .$$

From each set X_λ we pick up an arbitrary but fixed x_λ , from which we define the mapping

$$\beta : p \longmapsto \sum a_\lambda \square x_\lambda \quad . \quad [340]$$

Clearly, β is an R-hom. Besides, we have

$$(\mu \circ \beta)p = \mu(\sum a_\lambda \square x_\lambda) = \sum a_\lambda \square (\mu x_\lambda) = \sum a_\lambda \square (\alpha y_\lambda)$$

$$= \alpha(\sum a_\lambda \square y_\lambda) = \alpha p \quad [341]$$

which establishes [338].

||

(def) An exact sequence (of R-modules)

$$M \xrightarrow{\mu} N \longrightarrow 0 \qquad [342]$$

is said to *split* if

$$\exists \nu \in \mathrm{Hom}_R(N, M) : \mu \circ \nu = \hat{1}_N . \qquad [343]$$

Similarly, an exact sequence

$$0 \longrightarrow M \xrightarrow{\mu} N \qquad [344]$$

is *split* if [343] is satisfied.

Proposition XXXV

If an exact sequence (of R-modules)

$$M \xrightarrow{\mu} N \longrightarrow 0 \qquad [345]$$

is split, then

$$M \longleftrightarrow (\mathrm{im}.\mu \oplus \ker.\mu) \longleftrightarrow (N \oplus \ker.\mu) . \qquad [346]$$

Proof

Since [345] is split, we have

$$\exists\, \nu \in \mathrm{Hom}_R(N, M) : \mu \circ \nu = \hat{1}_N .$$

That μ is R-epi implies the following:

$$M/\ker.\mu \longleftrightarrow N = (\mu \circ \nu)\hat{1}_N = \mu(\mathrm{im}.\,\nu).$$

Hence [346] follows. ||

Proposition XXXVI

For a split exact sequence (of R-modules)

$$0 \longrightarrow M \xrightarrow{\mu} N , \tag{347}$$

if ν is any R-hom from N to M such that $\mu \circ \nu = \hat{1}_N$, then

$$M \longleftrightarrow (\mathrm{im}.\nu \oplus \ker.\nu) \longleftrightarrow (N \oplus \ker.\nu). \tag{348}$$

Proof

We leave the simple proof to the reader. ||

Proposition XXXVII

For an R-module P, the following conditions are equivalent

i) P is projective

ii) Every exact sequence of the form

$$M \xrightarrow{\mu} P \longrightarrow 0 \tag{349}$$

is split.

iii) P is a direct summand of a free R-module.

<u>Proof</u>

Our scheme of proof is : (i) $\Longrightarrow$ (ii) $\Longrightarrow$ (iii) $\Longrightarrow$ (i).

<u>Proof of (i) $\Longrightarrow$ (ii)</u>

Assume that P is projective. Hence, for every row-exact diagram

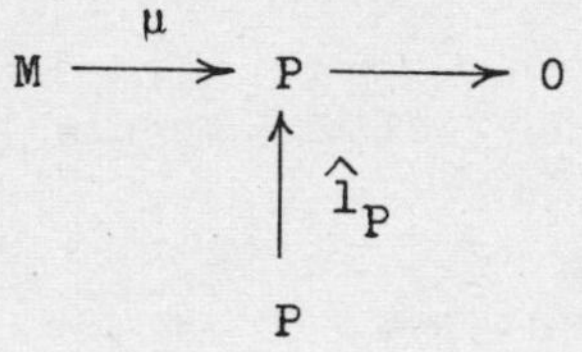

[350]

there exists $\nu \in \text{Hom}_R(P, M) : \mu \circ \nu = \hat{1}_P$. Therefore, by [342] the exact sequence [349] is split. ∎

<u>Proof of (ii) $\Longrightarrow$ (iii)</u>

Let F be a <u>free</u> R-module generated by P, then the following diagram is commutative (see [244]):

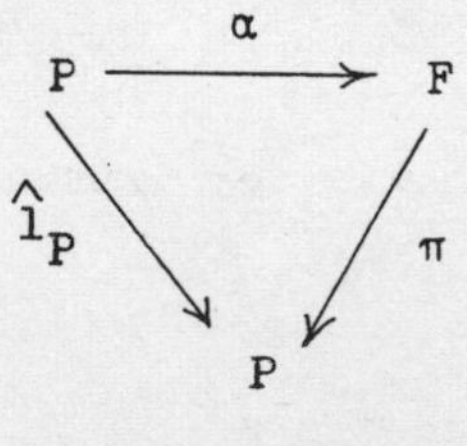

[351]

[351] says that π must be an R-epi, i.e. the following diagram is exact:

$$F \xrightarrow{\pi} P \longrightarrow 0 \qquad [352]$$

which must be split by the assumption (ii). Hence Proposition XXXV is applicable to give

$$F \longleftrightarrow P \oplus \ker.\pi \ . \qquad [353]$$

∎

<u>Proof of (iii) $\Longrightarrow$ (i)</u>

Let F be a <u>free</u> R-module such that

$$F = P \oplus Q \ . \qquad [354]$$

Define the <u>projection</u> and the <u>injection</u> mappings

$$\sigma : F \longrightarrow P \quad \text{and} \quad \eta : P \longrightarrow F \ . \qquad [355]$$

Consider a row-exact diagram:

$$\begin{array}{ccccc} M & \xrightarrow{\mu} & N & \longrightarrow & 0 \\ & & \uparrow{\scriptstyle \gamma} & & \\ & & P & & \\ & & \uparrow{\scriptstyle \sigma} & & \\ & & F & & \end{array} \qquad [356]$$

Since F is projective (in virtue of Proposition XXXIV) there exists

$\beta \in \mathrm{Hom}_R(F, M)$ to make the following diagram commutative:

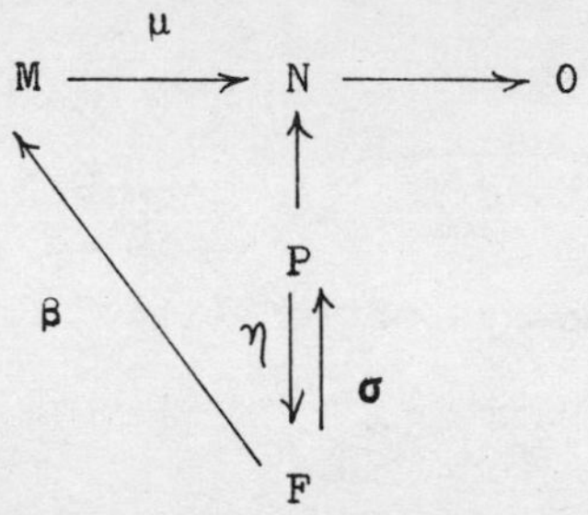

[357]

The R-hom $\beta \circ \eta$ now connects P to M. Consequently, P is projective. ||

(def) An R-module Q is said to be *injective* if, for every row-exact diagram

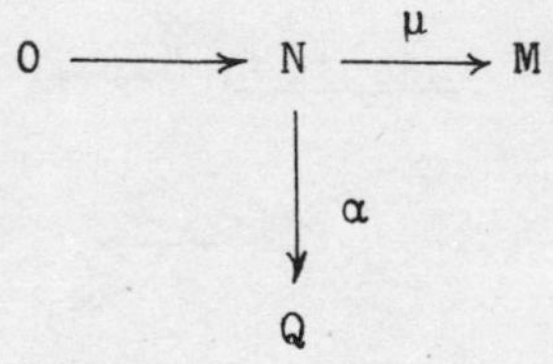

[358]

there exists $\beta \in \mathrm{Hom}_R(M, Q)$ such that the following diagram is commutative:

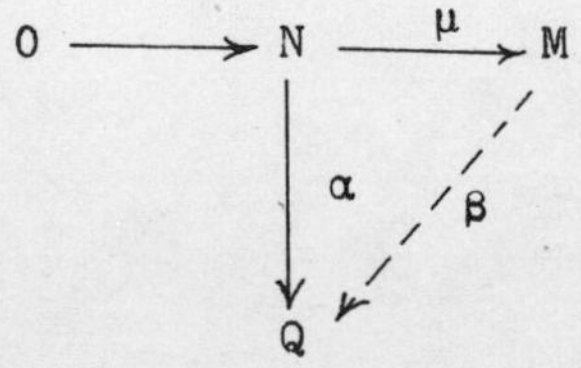

[359]

i.e. $\beta \circ \mu = \alpha.$ [360]

Clearly the concept of injective modules is dual to that of projective modules. Hence many properties of injective modules can be obtained by duality (i.e. arrow-reversing) from that of projective modules.

Now let us conclude this section by turning our attention to the notion of "factor-system" whose counter part in group theory is factor-set.

(def) For an extension (of R-modules)

$$0 \longrightarrow M \xrightarrow{\eta} E \xrightarrow{\beta} N \longrightarrow 0 , \qquad [361]$$

a mapping $\mu \in \mathrm{Map}(N, E)$ is called a representative mapping if

$$\beta \circ \mu = \hat{1}_N . \qquad [362]$$

For any $x \in N$, $\mu(x)$ is called a representative. It is important to emphasize that μ is not necessarily an R-hom. In other words, $\mu(x_1 + x_2) = \mu(x_2) + \mu(x_2)$ is incorrect in general.

In analogy to [383] of Chapter II (the multiplications there are now written additively here), for any fixed choice of μ in [361], we have

$$\mu(x_1) + \mu(x_2) = \mu(x_1 + x_2) \mod \eta M , \qquad [363]$$

for every $x_1, x_2 \in N$. Hence

$$\underset{1!}{\exists}\ f(x_1, x_2) \in M : \mu(x_1) + \mu(x_2) = \eta(f(x_1, x_2)) + \mu(x_1 + x_2) \quad [36$$

where $f \in \mathrm{Map}(N\ \ N, M)$ is called the first factor-set (of N i M) associated with μ. Further, for any $r \in R$ and $x \in N$,

$$\beta[\ \mu(r \,\square\, x) - r \,\square\, (\mu x)] = r \,\square\, x - \beta(r \,\square\, (\mu x)) = r \,\square\, x - r \,\square\, ((\beta\mu)x$$

$$= r \,\square\, x - r \,\square\, x = 0$$

or $$[\mu(r \,\square\, x) - r \,\square\, (\mu x)] \in \ker.\beta \ = \eta M \quad [365]$$

i.e. $$\underset{1!}{\exists}\ \tilde{f}(r, x) \in M : r \,\square\, (\mu x) = \eta(\tilde{f}(r, x)) + \mu(r \,\square\, x) \quad [366]$$

where $\tilde{f} \in \mathrm{Map}(R\ \ N, M)$ is called the second factor-set (of N in M) associated with μ. The pair $(f, \tilde{f})$ is then called a factor-system (of N in M) associated with μ. Clearly, to different choices of representative mappings there correspond different factor-systems. Let μ and μ' be representatives wi factor-systems $(f, \tilde{f})$ and $(f', \tilde{f}')$, then

$$\underset{1!}{\exists}\ h(x) \in M : \mu'(x) = \eta(h(x)) + \mu(x), \text{ for every } x \in N \quad [367]$$

where $f'(x_1, x_2)$ is defined by (similar to [364]):

$$\mu'(x_1) + \mu'(x_2) = \eta(f'(x_1, x_2)) + \mu'(x_1 + x_2). \quad [368]$$

By [367] and [364], we have

$$\mu'(x_1) + \mu'(x_2) = \eta(h(x_1) + h(x_2)) + \mu(x_1) + \mu'(x_2)$$

$$= \eta(h(x_1) + h(x_2) - h(x_1 + x_2)) + f(x_1, x_2) + \mu'(x_1 + x_2). \qquad [369]$$

A comparison between [368] and [369] yields

$$f'(x_1, x_2) - f(x_1, x_2) = h(x_1) + h(x_2) - h(x_1 + x_2).$$

Hence $f' - f$ measures the deviation of mapping h from a true <u>homomorphism</u> (of abelian groups). We now show that, together with $\tilde{f}' - \tilde{f}$, it detects the deviation of h from a true R-hom. From [366] and [367], we get

$$r \mathbin{\square} (\mu' x) = r \mathbin{\square} (\eta(h(x))) + r \mathbin{\square} (\mu x) = \eta(r \mathbin{\square} hx) + r \mathbin{\square} (\mu x)$$

$$= \eta(r \mathbin{\square} (hx)) + \eta(\tilde{f}(r, x)) + \mu(r \mathbin{\square} x)$$

$$= \eta(r \mathbin{\square} (hx)) + \eta(\tilde{f}(r, x)) + \mu'(r \mathbin{\square} x) - \eta(h(r \mathbin{\square} x)). \qquad [370]$$

Similar to [366], we have

$$r \mathbin{\square} (\mu' x) = \eta(\tilde{f}'(r, x)) + \mu(r \mathbin{\square} x)$$

When combined with [370] we obtain

$$\tilde{f}'(r, x) - \tilde{f}(r, x) = r \mathbin{\square} (hx) - h(r \mathbin{\square} x). \qquad [371]$$

Hence $\tilde{f}' - \tilde{f}$ vanishes if h is an R-hom.

Denote by δ and $\tilde{\delta}$ the mappings

$$(\delta h)(x_1, x_2) \equiv h(x_1) + h(x_2) - h(x_1 + x_2) \qquad [372]$$

and $(\tilde{\delta}h)(r, x) \equiv r \square (hx) - h(r \square x).$ [373

then, $$f' - f = \delta h \quad \text{and} \quad \widetilde{f'} - \widetilde{f} = \tilde{\delta}h$$ [374

which are analogous to [397] of Chapter II. Denote by $Fact_R(N, M)$ (cf. the notation following [400] of Chapter II) the set of all factor-systems of N in M. In analogy to [398] of Chapter II, we denote by $B_R^2(N, M)$ the set of all pairs $(\delta, \tilde{\delta})$. Hence

$$(f, \tilde{f}) = (f', \widetilde{f'}) \mod B_R^2(N, M).$$ [37

In this sense, a factor-system is uniquely determined by a giv extension modulo $B_R^2(N, M)$.

§. 4.11. Chain complexes, homology and cohomology modules.

In this section, R denotes a unitary ring. We shall call all unitary left R-modules simply "R-modules".

(def) A descending sequence $K \equiv \{K_n, \delta_n\}_{n \varepsilon I}$, of R-modules K_n and $\delta_n \varepsilon Hom_R(K_n, K_{n-1})$, is called a chain complex (of R-modules) if

$$\delta_n \circ \delta_{n+1} = 0, \text{ for every } n \varepsilon I,$$ [3

where 0 denotes the zero-mapping (i.e. $\mathrm{im}(\delta_n \circ \delta_{n-1}) = \{0\}$). The R-hom δ_n is called an *n-boundary operator*. $\ker.\delta_n$, to be denoted here by $Z_n(K)$, is called the *n-cycle module*; each element of $Z_n(K)$ is an *n-cycle*. $\mathrm{im}.\delta_{n+1}$, to be denoted here by $B_n(K)$ is called the *n-boundary module*; each element of $B_n(K)$ is an *n-boundary*. The *n-th homology module* of K is defined by

$$H_n(K) \equiv Z_n(K)/B_n(K) \ . \qquad [377]$$

An element of $H_n(K)$ is often referred to as a *homology class*. For simplicity of notation we often write δ in place of δ_n. Clearly the condition [376] is equivalent to

$$\mathrm{im}.\delta_{n+1} \subset \ker.\delta_n \ , \text{ for every } n \in I \ . \qquad [378]$$

A chain complex $K \equiv \{K_n, \delta_n\}$ has a diagrammatic form

$$K : \cdots \xrightarrow{\delta} K_{n+1} \xrightarrow{\delta} K_n \xrightarrow{\delta} K_{n-1} \xrightarrow{\delta} \cdots \ . \qquad [379]$$

(def) Two n-cycles, of a chain complex K, are said to be *homologous* if they belong to the same homology class.

Now, let us define the concept of the "cohomology" which was extensively discussed in Chapter II in the context of group theory.

(def) An ascending sequence $K \equiv \{K^n, \delta^n\}_{n \varepsilon I}$, of R-modules K^n and $\delta^n \varepsilon \mathrm{Hom}_R(K^n, K^{n+1})$, is called a cochain complex if

$$\delta^n \circ \delta^{n-1} = 0 \text{ , for every } n \varepsilon I \text{ .} \qquad [380]$$

Then δ^n is called an n-coboundary operator, $Z^n(K) \equiv \mathrm{ker}.\delta^n$ the n-cocycle module, $\mathrm{im}.\delta^{n-1} \equiv B^n(K)$ the n-coboundary module and

$$H^n(K) \equiv Z^n(K)/B^n(K)$$

the n-cohomology module of K. A cochain complex K has the form

$$K : \cdots \xleftarrow{\delta} K^{n+1} \xleftarrow{\delta} K^n \xleftarrow{\delta} K^{n-1} \xleftarrow{\delta} \cdots \qquad [381]$$

where we write δ in place of δ^n, $n \varepsilon I$, for simplicity. We see that [381] can be obtained from [379] by reversing the arrow and raising the indices. In other words, the duality carries all homological objects to cohomological objects and vice versa The duality is best seen in the following setting. Let us consider, for the moment, a unitary commutative ring R. Let M (with $\Omega_\square$) be an R-module and $K \equiv \{K_n, \delta_n\}$ a chain complex. Each $\mathrm{Hom}_R(K_n, M)$ is an abelian group; the addition is defined by

$$(f + f')x = f(x) + f'(x) \qquad [382$$

for every $x \varepsilon K_n$, f and $f' \varepsilon \mathrm{Hom}_R(K_n, M)$. Since R is commutat we can turn $\mathrm{Hom}_R(K_n, M)$ into an R-module by defining a composition:

$$\Omega_{\blacksquare}: R \times \mathrm{Hom}_R(K_n, M) \longrightarrow \mathrm{Hom}_R(K_n, M) \qquad [383]$$

with $$(r \blacksquare f)x = r \square (fx) , \qquad [384]$$

for every $r \in R$, $f \in \mathrm{Hom}_R(K_n, M)$ and $x \in K_n$. Consider the mapping

$$\delta^n : \mathrm{Hom}_R(K_n, M) \longrightarrow \mathrm{Hom}_R(K_{n+1}, M) \qquad [385]$$

with $$\delta^n : f_n \longmapsto (-)^{n+1} f_n \delta_{n+1} , \qquad [386]$$

for every $f_n \in \mathrm{Hom}_R(K_n, M)$. It is easy to see that δ^n is an R-hom (see the Problem Section). Furthermore, for any $f_{n-1} \in \mathrm{Hom}_R(K_{n-1}, M)$,

$$(\delta^n \circ \delta^{n-1})f_{n-1} = (-)^n \delta^n(f_{n-1}\delta_n) = (-)^n(-)^{n+1}(f_{n-1}\delta_n)\delta_{n+1}$$

$$= - f_{n-1}(\delta_n \circ \delta_{n+1}) = 0 . \qquad [387]$$

This shows that δ^n, $n \in I$, are n-coboundary operators. The ascending sequence

$$K^* : \cdots \xrightarrow{\delta^{n-2}} \mathrm{Hom}_R(K_{n-1}, M) \xrightarrow{\delta^{n-1}} \mathrm{Hom}_R(K_n, M) \xrightarrow{\delta^n} \cdots \qquad [388]$$

is consequently a cochain complex, to be denoted by $\mathrm{Hom}_R(K, M)$.

From this, the coboundary and cohomology are accordingly defined. The cohomology module, $H^n(K, M)$, of the cochain complex [388] is often referred to as the *cohomology of* K *with coefficients in* M. It is given by:

$$H^n(K, M) = Z^n(K^*)/B^n(K^*) \qquad [389]$$

where $\qquad K^* \equiv Hom_R(K, M)$.

For two chain complexes K and K' of R-modules (R is any unitary ring), a *chain homomorphism*, $f : K \longrightarrow K'$, is defined as a collection of R-hom $\{f_n\}_{n \in I}$ such that the following diagram is commutative:

$$\begin{array}{ccccccccc} \cdots & \longrightarrow & K_{n+1} & \xrightarrow{\delta_{n+1}} & K_n & \xrightarrow{\delta_n} & K_{n-1} & \xrightarrow{\delta_{n-1}} & \cdots \\ & & \downarrow f_{n+1} & & \downarrow f_n & & \downarrow f_{n-1} & & \\ \cdots & \longrightarrow & K'_{n+1} & \xrightarrow[\delta'_{n+1}]{} & K'_n & \xrightarrow[\delta'_n]{} & K'_{n-1} & \xrightarrow[\delta'_{n-1}]{} & \cdots \end{array} \qquad [390]$$

In other words,

$$f_{n-1}\, \delta_n = \delta'_n\, f_n \,, \quad n \in I \,. \qquad [391]$$

It is very tempting, at this point, to wander into the rich content of homology theory which, unfortunately, is not within our scope for a book of this size and of this nature. Therefor we close this section by the following easy property:

Proposition XXXVIII

Let $f : K \longrightarrow K'$ be a chain hom, then f induces some R-hom

$$f_{n*} : H_n(K) \longrightarrow H_n(K'), \quad \text{for } n \in I. \qquad [392]$$

Proof

Let notations be defined by diagram [390]. Commutativity of the diagramm implies

$$f_n \delta_{n+1} = \delta'_{n+1} f_{n+1} \qquad [393]$$

i.e. $f_n(\text{im. } \delta_{n+1}) = \delta'_{n+1}(\text{im.} f_{n+1}) \subset \delta'_{n+1} K'_{n+1} = \text{im.} \delta'_{n+1}$

or $$f_n : \text{im. } \delta_{n+1} \longrightarrow \text{im. } \delta'_{n+1} . \qquad [394]$$

Hence f_n induces the mapping

$$\bar{f}_n : K_n/\text{im. } \delta_{n+1} \longrightarrow K'_n/\text{ im. } \delta'_{n+1} \qquad [395]$$

with $\bar{f}_n : x \bmod (\text{im. } \delta_{n+1}) \longmapsto f_n(x) \bmod (\text{im. } \delta'_{n+1})$, [396]

for every $x \in K_n$.

Next, by $f_{n-1} \delta_n = \delta'_n f_n$, we conclude that

$$f_n : \text{ker. } \delta_n \longrightarrow \text{ker. } \delta'_n \ . \qquad [397]$$

Hence we can restrict $\overline{f}_n$ of [395] to the domain $H_n(K)$. This yields the sought-after induced R-hom

$$f_{n*} \equiv \overline{f}_n \Big|_{H_n(K)} \ . \qquad [398]$$

It is obvious that

$$f_{n*} : H_n(K) \longrightarrow H_n(K') \ . \qquad [399]$$

‖

Remark

[394] guarantees that the mapping [396] is well-defined. Then [397] guarantees that [399] has the correct codomain.

Problems with hints or solution for Chapter IV

R denotes an arbitrary ring unless otherwise stated.

Problem 1

Let R be a ring. If M is a simultaneous right and left R-module (with $\Omega_\square$) such that

$$r \square x = x \square r \qquad [P.1]$$

for every r, r' ε R and x ε M, show that

$$(rr') \square x = (r'r) \square x \qquad [P.2]$$

must be satisfied.

Hint for proof : It is similar to the remark following [11]. ∎

Problem 2

If M' is a submodule of a left R-module M, verify that M mod M' is a left R-module w.r.t. the composition $\Omega_\blacksquare$ defined by [52].

Proof

Consider any $r, r_1, r_2 \in R$ and $x, x_1, x_2 \in M$. Then

i) $r \blacksquare (\overline{x}_1 + \overline{x}_2) = r \blacksquare \overline{x_1 + x_2} = r \square (x_1 + x_2) \bmod M'$

$$= (r \square x_1 + r \square x_2) \bmod M'$$

$$= \overline{r \square x_1} + \overline{r \square x_2} = r \blacksquare \overline{x}_1 + r \blacksquare \overline{x}_2 \ . \ \blacksquare \qquad [P.3]$$

ii) $$(r_1 + r_2) \blacksquare \overline{x} = (r_1 + r_2) \square x \bmod M'$$

$$= \overline{r_1 \square x} + \overline{r_2 \square x} = r_1 \blacksquare x + r_2 \blacksquare x \ . \ \blacksquare \qquad [P.4]$$

iii) $$(r_1 r_2) \blacksquare \overline{x} = (r_1 r_2) \square x \bmod M'$$

$$= r_1 \square (r_2 \square x) \bmod M'$$

$$= r_1 \blacksquare \overline{r_2 \square x} = r_1 \blacksquare (r_2 \blacksquare x) \ . \ \blacksquare\blacksquare \qquad [P.5]$$

Problem 3

Let M be a left R-module. Show that

$$\text{ann.}M \mathrel{\vartriangleleft} R \ . \qquad [P.6]$$

Proof

ann.M is obviously a subring of R. It is also a right ideal of R since, for every $r \in R$ and $a \in \text{ann.}M$,

$$(ar) \square M = a \square (r \square M) \subset a \square M = \{0\} \qquad [P.7]$$

i.e. $ar \in \text{ann.}M$. ▮

Next, we have

$$(ra) \square M = r \square (a \square M) = r \square \{0\} = \{0\} \qquad [P.8]$$

i.e. $ra \in \text{ann}.M$. ||

Problem 4

Show that M as a left (R/ann.M)-module, as defined by [57], is faithful.

Proof

If r is an element of R such that

$$r \,\square\, x = 0 \;, \quad \text{for every} \quad x \in M \qquad [P.9]$$

then, by definition,

$$r \in \text{ann}.M \;.$$

But ann.M is the zero of the ring R/ann.M, hence M is faithful as a left (R/ann.M)-module. ||

Problem 5

Let M and M' be left R-modules. Show that $\text{Hom}_R(M, M')$ is a sub abelian group of Hom(M, M').

Proof

We shall follow the notation of [62] and [63].

Proposition V of Chapter II shows that Hom(M, M') is an abelian group with its composition Ω_* defined by, for every

$x \in M$ and $f_1, f_2 \in \mathrm{Hom}(M, M')$,

$$(f_1 * f_2)x = f_1x + f_2x \ . \qquad [P.10]$$

If $f_1, f_2 \in \mathrm{Hom}_R(M, M')$ then, for every $r \in R$,

$$\begin{aligned}(f_1 * f_2)(r \square x) &= f_1(r \square x) + f_2(r \square x) \\ &= r \blacksquare (f_1x) + r \blacksquare (f_2x) \\ &= r \blacksquare (f_1x + f_2x) \\ &= r \blacksquare ((f_1 * f_2)x) \ . \qquad [P.11]\end{aligned}$$

In other words,

$$(f_1 * f_2) \in \mathrm{Hom}_R(M, M') \ . \qquad [P.12]$$

Hence $\mathrm{Hom}_R(M, M')$ is a sub abelian group. ||

Problem 6

Show that I, as an I-module (cf. [153]), satisfies the ACC but not the DCC.

Proof

It is easy to see that I satisfies the ACC. To show that I does not satisfy the DCC let us consider any non-zero $x \in I$ Then we have the following descending chain,

$$\mathrm{Mod}((x)) \supset \mathrm{Mod}((x^2)) \supset \cdots , \qquad [P.13]$$

which does not terminate. ||

Problem 7

Show that the I-module $\mathbb{Q}_p$ defined in [154] to [157] satisfies the DCC but not the ACC.

Hint: One has simply to prove that any proper (additive) subgroup (of $\mathbb{Q}$) that contains $\mathbb{Q}$ must be of the form $p^{-n}\mathbb{Q}_p$ with integer n. ||

Problem 8

Show [167] of Proposition XII, for $n > m$.

Proof

The proof is similar to the case of $m \geq n$. For $n > m$, we have (cf. [165] of Proposition XII)

$$\begin{aligned} M_i &= M_i \cap (M_i + M') = M_i \cap (M_n + M') \\ &= M_i \cap M_n + M_i \cap M' = M_n + M_i \cap M' \\ &= M_n + M_n \cap M' = M_n . \end{aligned} \qquad [P.14]$$

Hence the ascending chain

$$M_1 \subset M_2 \subset \cdots$$

stops at a <u>finite</u> length n . ||

<u>Problem 9</u>

Complete the proof of Proposition XXII by showing that ann.x is modular.

<u>Proof</u>

Since we have shown

$$R \square x = M \text{ ,}$$ [P.1

it follows that

$$\exists t \in R : t \square x = x$$ [P.1

i.e. $r \square (t \square x) = r \square x$, for every $r \in R$

or $(rt) \square x - r \square x = 0$

or $(rt - r) \square x = 0.$

Hence, by definition,

$(rt - r) \in$ ann.x , for every $r \in R$ [P.1

i.e. ann.x is modular. ||

<u>Problem 10</u>

Let $\mathfrak{M}_R$ be the collection of all <u>simple</u> left R-modules. Let $\mathcal{H}_R$ be the collection of all modular maximal left ideals of R, then

$$M \in \mathfrak{M}_R \Longrightarrow \exists H \in \mathcal{H}_R : M \longleftrightarrow R/H \qquad [P.18]$$

and
$$H \in \mathcal{H}_R \Longrightarrow \exists M \in \mathfrak{M}_R : M \longleftrightarrow R/H . \qquad [P.19]$$

<u>Proofs</u>

<u>Proof of [P.18]</u>

From Propositions XXII and XXIII, we see that ann.M establishes the existence of the H of [P.18]. ||

The proof of [P.19] is left to the reader.

<u>Problem 11</u>

Let H be a left ideal of R, and let

$$(H : R) \equiv \{t \mid t \in R, tR \subset H\} \qquad [P.20]$$

then
$$\text{ann.}(R/H) = (H : R), \qquad [P. 21]$$

where R/H is considered as a left R-module.

<u>Proof</u>

R/H is a left R-module w.r.t. the composition (cf. [224]),

$$\Omega_* : R \times (R/H) \longrightarrow R/H , \text{ with}$$

$$\Omega_{\blacksquare} : (r', r \bmod H) \longmapsto r'r \bmod H , \qquad [P.22]$$

for every $r, r' \in R$. The condition $tR \subset H$, $t \in R$, of [P.20] can be written as

$$t \blacksquare (R \bmod H) = H = \text{the zero of } R/H . \qquad [P.23]$$

On the other hand ann.(R/H) is defined as

$$\text{ann.}(R/H) = \{ t \mid t \blacksquare (R/H) = \text{the zero of } R/H, \ t \in R \} . \qquad [P.24]$$

Comparison between [P.23] and [P.24] establishes the proposition.

Problem 12

Show [261] in the direction ($\Longleftarrow$).

Proof

By assumption, αH is a base of M. Then every $x \in M$ can be written uniquely in the form

$$x = \sum_{\lambda \in \Lambda} r_\lambda (\alpha h_\lambda) , \qquad [P.25]$$

for some $r_\lambda \in R$ and $h_\lambda \in H$. To show that $\{M, \alpha\}$ is free we have to verify, for any given left R-module W and $\beta \in \text{Map}(H, W)$, that

$$\exists_{1!} \mu \in \text{Hom}_R(M, W) : \mu \circ \alpha = \beta . \qquad [P.26]$$

The required μ is given by

$$\mu : x \longmapsto \sum_{\lambda \in \Lambda} r_\lambda (\beta h_\lambda) \qquad [P.27]$$

where r_λ and h_λ are defined by [P.25]. We leave to the reader to check that μ is an R-hom and is <u>unique</u> up to isomorphism, in the sense of [P.26]. ∎

<u>Problem 13</u>

Let M be a left R-module and

$$M = \bigoplus_{i=1}^{n} M_i \qquad [P.28]$$

where M_i are submodules of M. Define the mappings (i=1,..., n),

$$\pi_i : M \longrightarrow M_i ,$$

$$\text{by} \quad \pi_i : \sum_{j=1}^{n} x_j \longmapsto x_i \quad \text{where} \quad x_j \in M_j . \qquad [P.29]$$

Show that π_i (i=1, ..., n) are "projective" R-endomorphisms, i.e.

$$\pi_i \pi_j = \delta_{ij} \pi_i \qquad [P.30]$$

and such that

$$\sum_{i=1}^{n} \pi_i = \hat{1}_M \ . \quad (\hat{1}_M = \text{identity R-endo on M}) \qquad [P.31$$

Proof

It is obvious from the definition of π_i. The LHS of [P.31] has the meaning of

$$(\sum_i \pi_i)x = \sum_i \pi_i(x), \qquad [P.32$$

for every $x \varepsilon M$. ▮

Problem 14

Prove [309].

Proof

First we define the mappings

$$\sigma_i \equiv \pi_i \otimes \hat{1}_N \ , \quad i=1, \ldots, n \qquad [P.3$$

where π_i's are defined in the preceding problem, and $\hat{1}_N$ is the identity R-endo on N. Then

$$M \otimes_R N = \bigoplus_{i=1}^{n} \sigma_i (M \otimes_R N) \ . \qquad [P.3$$

Let β be the balanced mapping of [277]. We introduce now the restrictions,

$$\beta_i = \beta\Big|_{M_i \times N}, \quad i = 1, \ldots, n \; . \qquad [P.35]$$

The following mapping is also needed:

$$\pi_i \times \hat{1}_N : (x, y) \longmapsto (\pi_i x, y) \qquad [P.36]$$

for every $x \in M$ and $y \in N$.

Next, we want to show that $\sigma_i(M \otimes_R N)$ is precisely a tensor product of M_i and N. This leads, by Proposition XXIX, to the isomorphism between $\sigma_i(M \otimes_R N)$ and $M \otimes_R N$. Hence [309] is established in virtue of [P.34]. To show that $\sigma_i(M \otimes_R N)$ is a tensor product let us consider any abelian group G and any balanced mapping

$$\alpha_i : M_i \times N \longrightarrow G \; . \qquad [P.37]$$

Denote by γ_i the mapping $\alpha_i\,(\pi_i \times \hat{1}_N)$. Since

$$\gamma_i : (x, y) \longmapsto \alpha_i(\pi_i x, y), \quad x \in M \text{ and } y \in N, \qquad [P.38]$$

γ_i is clearly a balanced mapping from $M \times N$ to G. Hence by the definition of $M \otimes_R N$, we have

$$\exists_{1!}\, \mu_i \in \mathrm{Hom}(M \otimes_R N, G) : \mu_i \circ \beta = \gamma_i \; . \qquad [P.39]$$

In other words, the diagram

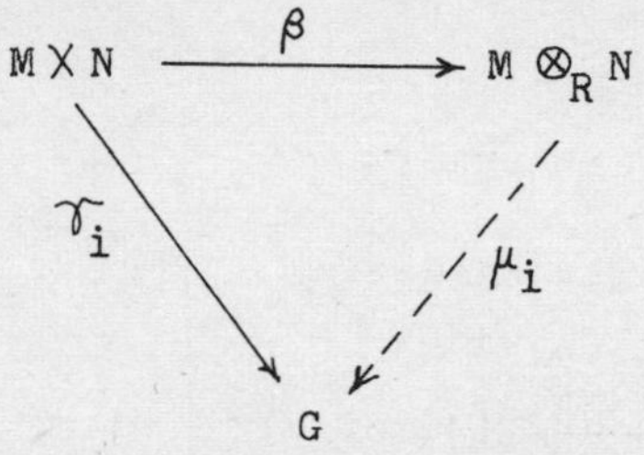

[P.40]

is commutative. Consider now the restriction

$$\mu_i\Big|_{\sigma_i(M \otimes_R N)} \equiv \nu_i \ . \qquad \text{[P.41]}$$

Then the following diagram

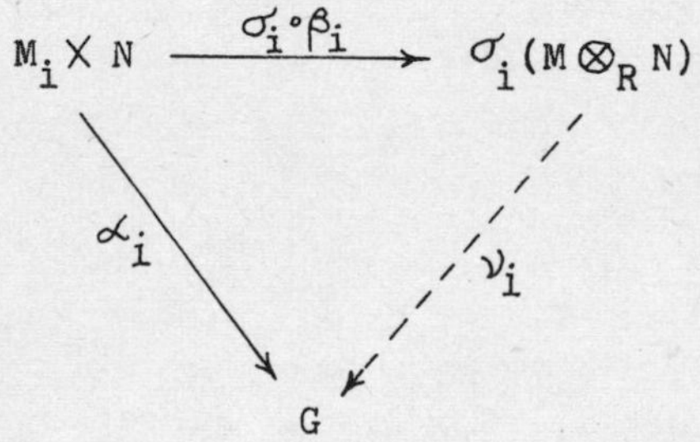

[P.42]

is commutative (the reader should check this by using the definitions of β_i, σ_i, ν_i and α_i ; the verification is very straightforward). Since G and α_i are both <u>arbitrary</u>, $\sigma_i(M \otimes_R N)$ is shown to be a tensor product of M_i and N. ∎

<u>Problem 15</u>

Prove Proposition XXXI in the "$\Longleftarrow$" direction.

Hint: By [321] one can show that im.α is a direct summand of L. ‖

Problem 16

Prove [329] of Proposition XXXII.

Hint: Chase the diagram along each square in [327]. ‖

Problem 17

Show that the mapping δ^n defined by [386] is an R-hom.

Proof

For any f_n, $f_n^{\cdot} \varepsilon \operatorname{Hom}_R(K_n, M)$, we have

$$\delta^n(f_n + f_n^{\cdot}) = (-)^{n+1}(f_n + f_n^{\cdot})\delta_{n+1}$$

$$= (-)^{n+1}(f_n \delta_{n+1} + f_n^{\cdot} \delta_{n+1})$$

$$= \delta^n f_n + \delta^n f_n^{\cdot} . \qquad\qquad [P.43]$$

Next, for every $x \varepsilon K_{n+1}$, we have

$$(\delta^n(r \blacksquare f_n))x = ((-)^{n+1}(r \blacksquare f_n)\delta_{n+1})x = (-)^{n+1}(r \blacksquare f_n)(\delta_{n+1} x)$$

$$= (-)^{n+1} r \square ((f_n \delta_{n+1})x) = r \square (((-)^{n+1} f_n \delta_{n+1})x)$$

$$= r \square ((\delta_n f_n)x)$$

i.e. $$\delta^n(r \blacksquare f_n) = r \square (\delta_n f_n) .$$ ‖ [P.4

CHAPTER V

On Module Homomorphisms

Summary

In this chapter we consider only modules over a division ring or a field. The topics covered, classic in nature, include: notion of dimensionality, simplicity, semi-simplicity, Fitting decompositions, minimal polynomials, eigen vectors, eigen values, characteristic polynomials, primary components and bilinear forms.

In this chapter Γ denotes a division ring, and K a field. All modules considered here are unitary left modules whose compositions are denoted by $\Omega_{\square}$ unless otherwise specified. K-modules are just vector spaces over K by definition.

§. 5.1. Dimensionality of a Γ-module

For convenience of discussion we shall use the conventional notion in set theory:

card.S $\equiv$ the cardinal number of the set S.

Our intention here is to discuss the concept of "dimension" (i.e. the "algebraic dimension") of a Γ-module. First we want to prove the following theorem:

Proposition I

Every non-zero Γ-module has a base.

Proof

The proof relies on Zorn's lemma. Let V be a non-zero Γ-module, then the collection L of all Γ-linearly independent subsets of V is non-empty (This is obviously not true if $V = \{0\}$). Let L be partially ordered by set-inclusion. For any chain, Σ (of L),

$$\Sigma : B_1 \subset B_2 \subset \cdots , \qquad [1]$$

with $B_i \in L$, we form the union

$$B \equiv \bigcup_i B_i . \qquad [2]$$

Since B is the union of $B_i \in \Sigma \subset L$, B is Γ-linearly independent. That is, $B \in L$. But $B_i \subset B$, for every index i, hence B is an upper bound of the chain Σ. Now all the conditions of Zorn's lemma are met (cf. Proposition X of Chapter III) therefore, by the lemma, there exists a maximal member, say M, of L. Since M is maximal as a Γ-linearly independent subset, it is a base of V. ||

The following theorem lays the foundation of "dimension" of a Γ-module.

Proposition II

All the bases of a given Γ-module have the same cardinal number.

Proof

Let X and Y be any two bases of a Γ-module V. Then there are three possibilities:

i) both card.X and card.Y are finite.

ii) card.X is finite but card.Y is infinite.

iii) both card.X and card.Y are infinite.

Case (i)

Let both $m \equiv$ card.X and $n \equiv$ card.Y be finite. To be specific we write

$$X \equiv \{x_1, \ldots, x_m\} \qquad \text{and} \qquad Y \equiv \{y_1, \ldots, y_n\}. \qquad [3]$$

Clearly, none of the x_i or y_i can be zero since X and Y, as bases of V, must be Γ-linearly independent. Consider first the element y_1. In terms of the base X we can write

$$y_1 \equiv \sum_{i=1}^{m} a_{1i} \,\square\, x_i \tag{4}$$

where $a_{1i} \varepsilon \Gamma$ $(i = 1, \ldots, m)$. The Γ-module composition for V is denoted by $\Omega_\square$. Since $y_1 \neq 0$, there exists at least one $a_{1j} \neq 0$. For convenience, let us assume that X is labelled such that $a_{11} \neq 0$ in [4]. Therefore $a_{11}^{-1} \varepsilon \Gamma$. From [4], we have

$$x_1 = a_{11}^{-1} \,\square\, y_1 - \sum_{i=2}^{m} (a_{11}^{-1} a_{1i}) \,\square\, x_i \;. \tag{5}$$

[5] states that x_1 is expressible in terms of the set

$$\{y_1, x_2, x_3, \ldots, x_m\} \equiv X_1 \;. \tag{6}$$

In other words, if the element x_1 in [3] is replaced by the <u>elements</u> in X_1 then the new set <u>generates</u> V. That is, the set

$$\{x_2, x_3, \ldots, x_m\} \cup X_1 \tag{7}$$

generates V. But [7] is just X_1 itself, hence X_1 generates V.

Now we can express y_2 in terms of X_1:

$$y_2 = a_{21}\square y_1 + \sum_{i=2}^{m} a_{2i}\square x_i \; . \qquad [8]$$

Clearly, not every a_{2i} $(i \neq j)$ is zero, otherwise it would lead to $y_2 = a_{21}\square y_1$ in contradiction to the Γ-linear independence of Y. Let us assume, by indexing X suitably, that $a_{22} \neq 0$ in [8]. Then a_{22}^{-1} exists in Γ. Consequently [8] yields

$$x_2 = -(a_{22}^{-1}a_{21})\square y_1 + a_{22}^{-1}\square y_2 - \sum_{i=3}^{m}(a_{22}^{-1}a_{2i})\square x_i \; . \qquad [9]$$

Hence x_2 is expressible in terms of the set

$$\{y_1, y_2, x_3, \ldots, x_m\} \equiv X_2 \; . \qquad [10]$$

By the same reasoning as before, we see that X_2 again generates V. If $n > m$ then this continuing process of replacing x_i by y_i leads to the conclusion that $X_m \equiv \{y_1, \ldots, y_m\}$ generates V. Therefore Y is Γ-linearly <u>dependent</u>. This is impossible since Y is a base. Therefore, we must have $n \leq m$. Exchanging the roles of X and Y we conclude that $m \leq n$. Consequently $m = n$. ▮

<u>Case (ii)</u>

Let $m = \text{card.}X$ be finite. By the process of [6], [10], etc., we see that $\text{card.}Y \leq m$. In other words $\text{card.}Y$ is finite. Hence this is reduced to case (i). Consequently, we have card.$Y = m$. ▮

Case (iii)

We assume that both card.X and card.Y are infinite. The proof, in this case, makes use of the following facts in set theory:

1) Let $U \equiv \bigcup_{\lambda \varepsilon \Lambda} S_\lambda$, where every set S_λ is finite. If the index set Λ is infinite then

$$\text{card.}U \leq \text{card.}\Lambda \quad . \qquad [11]$$

2) If two sets S and S' are such that card.S $\leq$ card.S' and card.S' $\leq$ card.S, then card.S = card.S' We note that card.S $\leq$ card.S' means that card.S is equal to the cardinal number of some subset of S'). This is the so-called "Schröder-Bernstein theorem".

Step 1

First, every $x \varepsilon X$ can be expressed as

$$x \equiv \sum_{\lambda \varepsilon \Lambda} r_\lambda \Box y_\lambda \text{ , with } y_\lambda \varepsilon Y \text{ and } r_\lambda \varepsilon K,\ r_\lambda = 0 \text{ p.p. } \lambda \varepsilon \Lambda. \qquad [12]$$

The sum of [12] has a finite number of terms due to the condition " $r_\lambda = 0$ p.p. $\lambda \varepsilon \Lambda$ ". We claim that each $y_\lambda \varepsilon Y$

must appear in the expression [12]for <u>some</u> $x \in X$. This is not difficult to see. Let $y \in Y$ be an element that does not appear in [12] for <u>all</u> $x \in X$. Since X is a base we can express y in terms of X whose elements are K-linear combinations of elements in $Y - \{y\}$ (i.e. $Y \cap_c \{y\}$). This means that Y is K-linearly dependent, contrary to the fact that Y is a base. Hence there is no such $y \in Y$ and our claim is established. ∎

<u>Step 2</u>

Consider a mapping f of Y into X, defined by

$$f : y_\lambda \longmapsto x \,, \quad \text{for} \quad y_\lambda \in Y \qquad [13]$$

where $x \in X$ is arbitrarily chosen (once and for all) such that y_λ appears in the expression for x. But x is expressed in terms of a <u>finite</u> number of elements from Y, due to the condition " $r_\lambda = 0$ p.p. $\lambda \in \Lambda$ ". Therefore, no matter how the mapping [13] is chosen, the sets

$$\overset{-1}{f}(x), \quad x \in \text{im}.f \,,$$

are <u>finite</u> and <u>disjoint</u>. It is obvious that $\overset{-1}{f}(x)$ <u>covers</u> Y, i.e.

$$\bigcup_{x \in \text{im}.f} \overset{-1}{f}(x) = Y \,. \qquad [14]$$

By [11] and [14], we conclude that

$$\text{card}.Y \leq \text{card}(\text{im}.f) \leq \text{card}.X \ . \qquad [15]$$

Exchanging the roles of X and Y, we have

$$\text{card}.X \leq \text{card}.Y \ . \qquad [16]$$

Finally, by Schröder-Berstein theorem, [15] and [16] lead to card.X = card.Y . ||

In view of the above theorem we now define the "dimension" of a Γ-module.

(def) The (algebraic) dimension of a Γ-module V is defined as the cardinal number of a base of V.

(notation) dim.V ≡ dimension of V .

Proposition III

Two Γ-modules are Γ-isomorphic iff they have the same dimension.

Proof

Necessity proof

Let f be a Γ-iso, from Γ-modules V to W. If B is a base of V, then every $x \in V$ has the form

$$x \equiv \sum_{\lambda \varepsilon \Lambda} k_\lambda \circ b_\lambda, \; b_\lambda \; \varepsilon \; B, \; k_\lambda \varepsilon \Gamma, \; k_\lambda = 0 \text{ p.p. } \lambda \varepsilon \Lambda . \qquad [17]$$

Thus

$$f(x) = \sum_{\lambda \varepsilon \Lambda} k_\lambda \circ (f b_\lambda), \quad k_\lambda = 0 \text{ p.p. } \lambda \varepsilon \Lambda$$

which implies that fB is a <u>base</u> of W. Since f is bijective, we have

$$\text{card}.B = \text{card}(fB)$$

i.e. $\dim.V = \dim.W$. ▮ [18]

<u>Sufficiency proof</u>

We leave this to the reader (or see the Problem Section). ▮▮

<u>Remark</u>

It is important to realize that the <u>zero</u> Γ-module $\{0\}$ has dimension <u>zero</u>, i.e.

$$\dim\{0\} = 0 , \qquad [19]$$

since the fact that $\{0\}$ has <u>no</u> base (by definition) leads to $\text{card}.\varnothing = 0$.

<u>Proposition IV</u>

Let W be a non-zero submodule of a Γ-module V . If U

is a set of generators of W, then U contains a base of W.

Proof

Step 1

First let us consider the case of a finite set U, and wri-

$$U \equiv \{u_1, \ldots, u_n\} . \qquad [20]$$

The theorem is obvious if U is Γ-linearly independent (then U is a base by definition). If U is Γ-linearly dependent then there is a set $\{a_i\}_{i=1,\ldots,n}$, $a_i \in \Gamma$, with at least one non-zero member, say a_t, such that

$$\sum_{i=1}^{n} a_i u_i = 0 . \qquad [21]$$

Hence we can write

$$u_t = -\sum_{i=1}^{n}{}' a_t^{-1} a_i u_i \qquad [22]$$

where the prime on the summation sign indicates $i \neq t$. In this way u_t is eliminated from [20]. Call the new set U. Clearly U' also generates W. If U' is still Γ-linearly dependent, then we repeat the elimination process on U'. Since U is a finite set, this procedure ends eventually to yield a Γ-linearly independent set, hence a base of W. ∎

Step 2

Now let us consider the case of an *infinite* dimensional V and an *infinite* set U. The proof is similar to that of Proposition I. Since W is non-zero, the collection L of all Γ-linearly independent subsets of U is non-empty. Let L be partially ordered by set inclusion. For any chain Σ of L,

$$\Sigma : B_1 \subset B_2 \subset \cdots , \qquad [23]$$

with $B_i \in L$, we can form $B \equiv \bigcup_i B_i$. Then, following the same argument of Proposition I, we see that B is an upper bound of the chain Σ. Therefore, Zorn's lemma is applicable. Hence L has a *maximal* member, say M, as a Γ-linearly independent subset of U. Consequently, M is a *base* of W. ||

Remark

The above theorem has the important special case: If U is a set of generators of V then U contains a base of V.

Proposition V

Let W be a Γ-linearly independent subset of a non-zero Γ-module V. Then, for any given base X of V, there is a subset of X such that its union with W is again a base of V.

Proof

Clearly $W \cup X$ is a set of generators of V. First, if V

is finite-dimensional, we can remove some elements of X, one at a time, from the set $W \cup X$ (as in Step 1, of the preceding theorem) until the remaining set just becomes Γ-linearly independent. Hence we have a base (of V) which is the union of W with a subset of X.

Next, if V is infinite-dimensional, then we can remove some elements of X from $W \cup X$, as in Step 2 of the preceding theorem until we get a base of V. ||

Proposition VI

Let V be a finite-dimensional Γ-module, then

1) If U is a proper submodule of V, then $\dim.U < \dim.V$.

2) If U_1 and U_2 are submodules of V, then

$$\dim(U_1 + U_2) + \dim(U_1 \cap U_2) = \dim.U_1 + \dim.U_2 . \qquad [24$$

3) $\dim(U_1 \oplus U_2) = \dim.U_1 + \dim.U_2$. [25

4) $\dim(V/U) = \dim.V - \dim.U$. [26

Proof

We leave the proofs of (1) and (2) to the reader (or see the Problem Section). To prove (3), we only have to use [24]. Since $U_1 \oplus U_2$ implies that $U_1 \cap U_2 = \{0\}$, we have

$$\dim(U_1 \cap U_2) = \dim\{0\} = 0$$

in view of [19]. Hence [24] yields [25]. ▮

To prove [26] let us choose bases X and Y, for V and U respectively. By Proposition V, there exists a subset X' of X such that X' ∪ Y is a base of V. Denote by W the Γ-module generated by X'. Then

$$V = U \oplus W. \qquad [27]$$

But the mapping

$$(u + w) \bmod U \longmapsto w, \quad \text{for every } u \in U \text{ and } w \in W,$$

establishes a Γ-iso between V/U and W. Therefore, [26] follows immediately from [27]. ▮▮

For a finite-dimensional Γ-module, the concept of "codimension" is a useful one:

(def) Let U be a submodule of a Γ-module V, then the codimension of U is defined as the dimension of a direct supplement of U. Denote by codim.U the codimension of U, then

$$\text{codim.}U = \dim.V - \dim.U .$$

§. 5.2. Simplicity and semi-simplicity of a Γ-endo.

In this section and for the remainder of the chapter, all the Γ-modules are assumed to be finite-dimensional. V denotes a Γ-module (by this we mean a unitary left Γ-module as stated in the beginning of the chapter).

Let F be a subset of $\mathrm{End}_{\Gamma}V$ in the following definitions

(def) A submodule U of V is said to be F-invariant (i.e. invariant under F) if

$$fU \underset{\text{set}}{\subset} U , \quad \text{for every} \quad f \in F . \qquad [28]$$

If $fU = U$, for every $f \in F$, then we say that U is F-stable (i.e. stable under F).

(def) A non-zero Γ-module V is said to be F-simple (i.e. F-irreducible) if V has no non-zero F-invariant submodules. Otherwise, V is said to be F-non-simple (i.e. F-reducible).

(def) A Γ-module V is said to be F-semi-simple (i.e. F-completely reducible) if every F-invariant submodules of V has an F-invariant direct supplement. It is clear that

$$\text{F-simple} \underset{\not\Longleftarrow}{\Longrightarrow} \text{F-semi-simple}. \qquad [29]$$

(def) A Γ-module V is said to be F-indecomposable if V is not expressible as a direct sum of non-zero F-invariant submodules of V . Otherwise V is F-*decomposable*.

It is obvious that

$$\text{F-simple} \;\underset{\not\Longleftarrow}{\Longrightarrow}\; \text{F-indecomposable.} \qquad [30]$$

The above definitions can be used to describe Γ-endomorphisms:

(def) A subset F of End V is said to be *simple* (on V) if V is F-simple. Otherwise F is *non-simple*.

(def) A subset F of End V is said to be *semi-simple* (on V) if V is F-semi-simple. Similarly, F is *indecomposable* if V is F-indecomposable.

(def) Two Γ-endos f and f' are said to *commute* with each other if

$$f \circ f' = f' \circ f . \qquad [31]$$

Proposition VII

Every F-invariant submodule of a F-semi-simple Γ-module is F-semi-simple.

Proof

The proof patterns after that of Proposition XXIV of Chapter V. Let V be an F-semi-simple Γ-module, and let V' be an

F-invariant submodule of V . Then

$$V = V' \oplus V''$$

where V" is also an F-invariant submodule of V . Let W be any F-invariant submodule of V" then, by nature of V , we have

$$V = W \oplus W' \quad [32]$$

where W' is F-invariant. Consequently,

$$V' = V' \cap V = V' \cap (W \oplus W')$$

i.e. $$V' = W \oplus (V' \cap W') . \quad [33]$$

Obviously, $(V' \cap W')$ is F-invariant since V' is. Therefore V' is F-semi-simple. ||

Proposition VIII

A subset F of $\mathrm{End}_{\Gamma}V$ is semi-simple and non-simple iff V is expressible as a direct sum of F-invariant F-simple submodules.

Proof

The necessity proof

Since F is semi-simple and non-simple, V is F-semi-simple and F-non-simple. Therefore, by definition, V has at least one non-zero F-invariant submodule that has an F-invariant direct supplement, i.e.

$$V = V' \oplus U$$

where V' and U are non-zero F-invariant submodules of V. If V' is F-non-simple then, by definition, V' has a non-zero F-invariant submodule, say V_1'. If V_1' is still F-non-simple then V_1' has a non-zero F-invariant submodule V_2'. Repeating this argument we see that V' must have an F-simple F-invariant submodule, say V_1. Since V_1 is also a submodule of V and since V is F-semi-simple, V_1 has an F-invariant direct supplement, say V''. That is

$$V = V_1 \oplus V'' . \qquad [34]$$

Since V'' is F-semi-simple, by Proposition VII, we can push the above argument on V'' (in place of V) until we reach the stage (since V is finite-dimensional):

$$V = V_1 \oplus V_2 \oplus \cdots \oplus V_m \qquad [35]$$

where all V_i are F-simple and F-invariant. ▮

The sufficiency proof

Now we assume, by the condition of the proposition,

$$V = \bigoplus_{i=1}^{m} V_i \qquad [36]$$

where each V_i is F-invariant and F-simple. Consider any F-invariant proper submodule U, of V. Since $U \neq V$, we have

$$\exists j (1 \leq j \leq m) : U \cap V_j \neq V_j . \qquad [37]$$

For convenience, let us assume that we have labelled this V_j to be V_1 in [36]. Then [37] is rewritten as :

$$U \cap V_1 \neq V_1 \qquad [38]$$

It is easy to show (we leave this to the reader; or see the Problem Section) that

$$U \cap V_1 = \{0\} . \qquad [39]$$

Define $U_1 \equiv U + V_1$. Then we can write, by [39],

$$U_1 \equiv U \oplus V_1 . \qquad [40]$$

U_1 is obviously F-invariant since both U and V_1 are. Besides V_1 is F-simple. If the new F-invariant submodule U_1 of [40] is still a proper submodule of V, then we can push the above argument on U_1 (in place of U) to reach a new submodule

$$U_2 = U_1 \oplus V_2 = U \oplus V_1 \oplus V_2 \qquad [41]$$

where V_2 is such that $U_1 \cap V_2 = \{0\}$ with the same labelling assumption as V_1. This process can continue until we reach a stage when $U_r = V$, for some integer $r < m$ (m is <u>finite</u> since V is finite-dimensional). Consequently, we have

$$V = U \oplus V_1 \oplus \cdots \oplus V_r \qquad [42]$$

Denote by W the sum $\bigoplus_{i=1}^{r} V_r$. Since W is clearly F-invariant, it is the required direct supplement of U. Thus V is both F-semi-simple and F-non-simple. ▮

§. 5.3. <u>Projection mapping and ring of Γ-endos</u>.

Projection mappings are very useful tools in dealing with Γ-modules. Throughout this section, V denotes a Γ-module.

(def) Let $V' \oplus V''$ be a direct-sum decomposition of V, then the Γ-endo defined by

$$\pi : V \longrightarrow V'$$

with $$\pi : v' + v'' \longmapsto v', \qquad [43]$$

for every $v' \in V'$ and $v'' \in V''$, is called a *projection mapping* from V onto V'.

<u>Proposition IX</u>

A Γ-endo π on V is a projection mapping iff $\pi^2 = \pi$.

Proof

Necessity proof

Let $V = V' \oplus V''$ be a direct-sum decomposition. Let π be a projection mapping from V onto V'. For every $v \in V$, we write $v = v' + v''$, by the decomposition. Then

$$\pi^2 v = \pi^2(v' + v'') = \pi v' = v' = \pi v .$$

i.e. $$\pi^2 = \pi . \quad \blacksquare \qquad [44]$$

Sufficiency proof

Assume now $\pi^2 = \pi$. Let $V' \equiv \text{im}.\pi$ and $V'' \equiv \text{ker}.\pi$, then we can show that

$$V = \text{im}.\pi \oplus \text{ker}.\pi = \pi V \oplus (\hat{1}_V - \pi) V . \qquad [45]$$

Step 1

For every $v \in V$, we have

$$v = \pi v + (v - \pi v) = \pi v + (\hat{1}_V - \pi) v$$

i.e. $$V = \pi V + (\hat{1}_V - \pi) V = \text{im}.\pi + (\hat{1}_V - \pi) V . \qquad [46]$$

It is not difficult to see that $(\hat{1}_V - \pi)V = \text{ker}.\pi$. First, for any $z \in \text{ker}.\pi$,

$$\pi z = 0 = z - z = \hat{1}_V z - z$$

i.e. $$z = (\hat{1}_V - \pi) z \ \varepsilon\ (\hat{1}_V - \pi) V$$

or $$\ker.\pi \subset (\hat{1}_V - \pi) V . \qquad [47]$$

Next, for every $v \ \varepsilon\ V$, we have

$$\pi((\hat{1}_V - \pi)v) = \pi v - \pi^2 v = \pi v - \pi v = 0$$

i.e. $$(\hat{1}_V - \pi) V \subset \ker.\pi . \qquad [48]$$

[47] and [48] imply that

$$\ker.\pi = (\hat{1}_V - \pi) V . \qquad [49]$$

Hence $$V = \text{im}.\pi + \ker.\pi . \blacksquare \qquad [50]$$

<u>Step 2</u>

It remains to show that [50] is a <u>direct</u> sum. Consider any $x \ \varepsilon (\text{im}.\pi \cap \ker.\pi)$. Then x has the form $x \equiv \pi v$, for some $v \ \varepsilon\ V$. Since $x \ \varepsilon\ \ker.\pi$, we have $\pi x = 0$. Hence

$$x = \pi v = \pi^2 v = \pi x = 0 \qquad [51]$$

Hence $$\text{im}.\pi \cap \ker.\pi = \{0\} . \blacksquare \qquad [52]$$

By the direct-sum decomposition [45], we can write every $v \ \varepsilon\ V$ into the form

$$v = v' + v'', \quad v' \in \pi V, \quad v'' \in \ker.\pi \ .$$

Then, it is obvious that

$$\pi : v' + v'' \longmapsto v' \ .$$

Thus, π is a projection mapping. ||

Proposition X

A subset F of $\mathrm{End}_{\Gamma}V$ is decomposable iff there exists a projective mapping, not being $\hat{0}_V$ or $\hat{1}_V$, that commutes with each member of F.

Proof

We give here only the "necessity" proof. The "sufficiency" proof is left to the reader (or see the Problem Section).

Necessity Proof

Since the decomposability of F implies the F-decomposability of V, we have a direct sum

$$V = \bigoplus_{i=1}^{m} V_i \qquad (m > 1) \qquad [53]$$

where each V_i is a non-zero F-invariant submodule of V. Let π_i be the projection mappings from V onto V_i, then π_i are obvious neither $\hat{0}_V$ nor $\hat{1}_V$ since each V_i is non-zero. By [53] we can write every $v \in V$ into

$$v \equiv \sum_{i=1}^{m} v_i \quad \text{with} \quad v_i \in V_i .$$

Consider any $f \in F$, then

$$fv = f(\sum_{i=1}^{m} v_i) = \sum_{i=1}^{m} fv_i$$

where $fv_i \in V_i$ since V_i are F-invariant. Therefore,

$$(\pi_j \circ f)v = \sum_{i=1}^{m} \pi_j(fv_i) = \sum_{i=1}^{m} \delta_{ij}(fv_i) = fv_j = (f \circ \pi_j)v .$$

Hence $\pi_j \circ f = f \circ \pi_j$. ||

For a Γ-module V, we have seen (Proposition II of Chapter IV) that $End_\Gamma V$ is a ring. In fact it is a "Von Neumann regular ring", to be defined below.

(def) A ring R is called a Von Neumann regular ring if

$$\exists t \in R : rtr = r, \quad \text{for every } r \in R . \qquad [54]$$

Proposition XI

For a unitary ring R, the following conditions are

equivalent:

1) R is a Von Neumann regular ring.

2) For every given $r \in R$, the left ideal Rr is generated by a idempotent. Similarly, the right ideal rR is also generat by an idempotent.

3) For every given $r \in R$,

$$\exists r' \in R : R = Rr \oplus Rr' . \qquad [55]$$

Proof

Our scheme of proof is: (1) $\Longrightarrow$ (2) $\Longrightarrow$ (3) $\Longrightarrow$ (1) .

Proof of (1) $\Longrightarrow$ (2) :

From [54], we have

$$(tr)^2 = trtr = t(rtr) = tr, \qquad [56]$$

hence tr is an idempotent. Clearly tr generates Rr (why? Se the Problem Section).

Similarly,

$$(rt)^2 = (rtr)t = rt ,$$

hence rt is the idempotent that generates rR . ▮

Proof of (2) $\Longrightarrow$ (3) :

We shall prove the case of the <u>left</u> ideal Rr since the proof for the <u>right</u> ideal rR is entirely similar. Let $r \in R$. Assume that e is an idempotent that generates Rr. Then we have

$$R = R1 = R(e + 1 - e) = Re + R(1 - e) . \qquad [57]$$

It is easy to see that the sum on the RHS of [57] is a <u>direct sum</u> Let $x \in Re \cap R(1 - e)$, then

$$x = ye , \quad \text{for some} \quad y \in R \qquad [58]$$

and $$x = z(1 - e), \quad \text{for some} \quad z \in R . \qquad [59]$$

Thus, using the idempotent property $e^2 = e$, we get

$$x = ye = ye^2 = (z(1 - e))e = z(e - e^2) = z0 = 0 .$$

In other words,

$$Re \cap R(1 - e) = \{0\} . \qquad [60]$$

Hence $$R = Re \oplus R(1 - e) . \qquad [61]$$

But e generates Rr by assumption, i.e. Rr = Re , therefore [61] becomes:

$$R = Rr \oplus R(1 - e) \qquad [62]$$

which is [55]. ▮

Proof of (3) $\Longrightarrow$ (1) :

By assumption we have, for any given $r \in R$,

$$\exists r' \in R : R = Rr \oplus Rr' . \quad [63]$$

Therefore we can decompose 1 into:

$$1 = xr + yr' , \quad \text{for some } x, y \in R \quad [64]$$

i.e. $$r = r(xr + yr') = rxr + ryr' . \quad [65]$$

Since [63] is a direct sum of ideals (cf. [63] and [203] of Chapter III), we have

$$(Rr)(Rr') = \{0\} \quad [66]$$

which implies that $(1 r)(yr') = 0$, i.e. $ryr' = 0$. Consequentl [65] becomes

$$r = rxr . \quad [67]$$

Since r is arbitrary, R is a Von Neumann regular ring by definition. ||

Proposition XII

Let V be a Γ-module, then $End_{\Gamma}V$ is a Von Neumann regul ring.

Proof

For convenience, we write $\text{End}_{\Gamma}V = W$. For any $f \in W$, we define the <u>projection</u> mapping from V to fV:

$$\pi : V \longrightarrow fV . \qquad [68]$$

Then, by [45] of Proposition IX,

$$V = \text{im.}\pi \oplus \text{ker.}\pi = fV \oplus \text{ker.}\pi . \qquad [69]$$

Now let us introduce a Γ-endo f^*. For any $v \in V$, we can write (by [69])

$$v \equiv fv' + v'' , \qquad [70]$$

with $v' \in V$ and $v'' \in \text{ker.}\pi$. Define the mapping f^* by

$$f^* : (fv' + v'') \longmapsto v' . \qquad [71]$$

Then it is easy to see that $f \circ f^* = \pi$. First, by the definition of f^*,

$$(f \circ f^*) : \text{ker.}\pi \longrightarrow \{0\} . \qquad [72]$$

Next, for any $v \in V$,

$$(f \circ f^*)(fv) = f(f^*(fv)) = f(v) = \pi(fv)$$

i.e. $$(f \circ f^*)\Big|_{fV} = \pi\Big|_{fV} . \qquad [73]$$

[72] and [73] imply that

$$f \circ f^* = \pi \qquad [74]$$

or $$\pi \circ W = f \circ f^* \circ W \subset f \circ W . \qquad [75]$$

On the other hand, we have

$$fv = \pi v = \pi^2 v = \pi(\pi v) = \pi fv$$

i.e. $$f = \pi \circ f$$

or $$f \circ W = \pi \circ f \circ W \subset \pi \circ W . \qquad [76]$$

By [75] and [76] we establish that $f \circ W = \pi \circ W$. In other words, for every given $f \in W$, the right ideal fW is generated by the idempotent π (since $\pi^2 = \pi$). By (2) of Proposition XI, we conclude that W is a Von Neumann regular ring. ||

§. 5.4. Fitting decomposition w.r.t. a Γ-endo.

A very useful decomposition, though not as refined as some other decompositions, is the so-called "Fitting decomposition" of a Γ-module w.r.t. one of its Γ-endomorphisms. A Fitting decomposition decomposes a Γ-module into two components: the "nil-component" and the "iso-component". Throughout this section

V denotes a Γ-module.

(def) A Γ-endo f, on V, is said to be non-singular if f has an inverse, i.e. if

$$\exists f' \in \mathrm{End}_{\Gamma} V : f' \circ f = \hat{1}_V \qquad [77]$$

where $\hat{1}_V$ is the identity Γ-endo on V. Otherwise, we say that f is singular.

(def) A Γ-endo f, on V, is said to be nilpotent if

$$\exists m \in I_+ : f^m = \underbrace{f \circ \cdots \circ f}_{m \text{ copies}} = \hat{0}_V \qquad [78]$$

where $\hat{0}_V$ is the zero Γ-endo on V.

Proposition XIII

Let f be a Γ-endo on V, then

$$f \text{ is non-singular} \iff f \text{ is a } \Gamma\text{-iso on } V. \qquad [79]$$

(We leave the proof to the reader).

(def) Let f be a Γ-endo on V then the Fitting nil-component (or simply the "nil-component") of V, w.r.t. f, is defined by:

$$V_{f,o} \equiv \{z \mid z \in V, (f)^i z = 0, \text{ for some } i \in I_+\} \qquad [80]$$

(def) The Fitting iso-component (or simply "iso-component") of V, w.r.t. $f \in \mathrm{End}_{\Gamma} V$, is defined by:

$$V_{f,1} \equiv \bigcap_{i=1}^{\infty} (f^i V) \qquad [81]$$

Proposition XIV

For any Γ-endo f on V, the Fitting components of V w.r. f have the following properties:

i) $V_{f,o}$ and $V_{f,1}$ are submodules of V. [82]

ii) Both $V_{f,o}$ and $V_{f,1}$ are f-invariant, in fact

$$f\, V_{f,o} \subset V_{f,o} \quad \text{and} \quad fV_{f,1} = V_{f,1}\,. \qquad [83]$$

iii) $f\big|_{V_{f,o}}$ is nilpotent and $f\big|_{V_{f,1}}$ is a Γ-iso. [84]

Proof

Proof of (i) It is obvious that

$$V \underset{\Gamma\text{-mod}}{\supset} fV \underset{\Gamma\text{-mod}}{\supset} f^2V \supset \cdots . \qquad [85]$$

Denote the above descending chain by Σ_V. Then there are two possibilities: either Σ_V ends at $\{0\}$ after a finite length, or it becomes f-<u>stable</u> after a finite length r, without terminating at $\{0\}$. In the first case we have

$$V \supset fV \supset \cdots \supset \{0\} . \qquad [86]$$

Therefore, by the definitions,

$$V_{f,1} = \{0\} \quad \text{and} \quad V_{f,o} = V \qquad [87]$$

which give the trivial decomposition $V = V \oplus \{0\}$. In the second case, i.e. if Σ_V becomes f-<u>stable</u> after a finite length r without terminating at $\{0\}$, then we have

$$V \supset fV \supset \cdots \supset f^r V = f^{r+1} V = \cdots . \qquad [88]$$

Hence $$V_{f,1} \equiv \bigcap_{i=1}^{\infty} f^i V = f^r V . \qquad [89]$$

It is easy to verify that $V_{f,1}$ is a submodule of V since, for example,

$$f^r(x+y) \in f^r V = V_{f,1} , \quad \text{for every } x, y \in V_{f,1} . \qquad [90]$$

Other axioms for a Γ-module are also trivially satisfied by $V_{f,1}$.

Next, consider the Γ-modules

$$U_i \equiv \{ x \mid x \in V , f^i x = 0 \} , \quad i \in I_+ . \qquad [91]$$

Then

$$U_1 \subset U_2 \subset \cdots \qquad [92]$$

and $\exists\, s \in I_+ : U_s = U_{s+1} = \cdots .$ [93]

Therefore,

$$V_{f,o} = U_s \ . \qquad [94]$$

It is trivial to see that $V_{f,o}$ is a submodule of V. ▮

Proof of (ii). Obvious.

Proof of (iii). Since $V_{f,o} = U_s$ we have

$$f^s(V_{f,o}) = f^s U_s = \{0\} \qquad [95]$$

i.e. $f\big|_{V_{f,0}}$ is nilpotent. ▮

It is also obvious that $f\big|_{V_{f,1}}$ is a Γ-iso since

$$V_{f,1} = f(V_{f,1}), \qquad [96]$$

in virtue of [88] and [90]. ▮▮

Proposition XV

For any $f \in \mathrm{End}_\Gamma V$,

$$V = V_{f,o} \oplus V_{f,1} \quad . \qquad [97]$$

Proof

Step 1

We want to show that

$$V = V_{f,o} + V_{f,1} \quad . \qquad [98]$$

Let m be the larger number of r and s which are given by [88] and [93]. Since $m \geq r$ we have, by [88],

$$f^{2m}V = f^{m}V \qquad [99]$$

which implies that, for every $v \in V$,

$$\exists_{1!} v' \in V : f^{2m}v' = f^{m}v \qquad [100]$$

or $\quad f^{m}(v - f^{m}v') = 0$

i.e. $\quad (v - f^{m}v') \in V_{f,o} \quad .$

Hence,

$$v = (v - f^{m}v') + f^{m}v' \in (V_{f,o} + V_{f,1}) .$$

i.e. $\quad V = V_{f,o} + V_{f,1} \quad . \; ▮ \qquad [101]$

Step 2

It remains to show that [101] is actually a direct sum.

Since $m \geq s$, we have

$$f^m : V_{f,o} \longrightarrow \{0\}$$ [102

whose restriction (to be denoted by f^{m*})

$$f^{m*}: (V_{f,o} \cap V_{f,1}) \longrightarrow \{0\}$$

is a Γ-mono by nature of $V_{f,1}$. Therefore

$$V_{f,o} \cap V_{f,1} = \{0\}$$ [103

i.e. $V = V_{f,o} \oplus V_{f,1}$. ||

(def) A bracket product between two elements $f, f' \in \operatorname{End} V$ is defined by the difference of composite mappings, i.e.

$$f \circ f' - f' \circ f \equiv [f, f'] .$$ [104

Hence f commutes with f' if $[f, f'] = \hat{0}_V$.

Proposition XVI

If f and g are two Γ-endos on V, then

$$f \circ g^r = \sum_{i=0}^{r} (-)^i {}_rC_i \, g^{r-i} \circ f^{(i)}$$ [10

and $$g^r \circ f = \sum_{i=0}^{r} {}_rC_i\, f^{(i)} \circ g^{r-i} \qquad [106]$$

where $f^{(i)} \equiv [g, f^{(i-1)}]$ for $i > 0$, $f^{(o)} \equiv f$ and $g^o \equiv \hat{1}_V$.

Proof

We shall only prove [105] since the proof for [106] is similar. The proof is to be carried out by mathematical induction. For economy of notation we shall write ab in place of $a \circ b$.

First, [105] is obviously true for $r = 0$ since

$$f = (-)^o\, {}_oC_o\, g^o\, f^{(o)} = \hat{1}_V\, f = f\ . \qquad [107]$$

Next, by the induction hypothesis, let us assume that [105] is true for $r = m - 1$. Then

$$f\, g^{m-1} = \sum_{i=0}^{m-1} (-)^i\, {}_{m-1}C_i\, g^{m-i-1}\, f^{(i)}$$

or $$f\, g^m = \sum_{i=0}^{m-1} (-)^i\, {}_{m-1}C_i\, g^{m-i-1}(f^{(i)}g)\ . \qquad [108]$$

But, by definition,

$$f^{(i)}g = g\,f^{(i)} - [g, f^{(i)}] = g\,f^{(i)} - f^{(i+1)} \qquad [109]$$

thus [108] becomes

$$f\,g^m = \sum_{i=0}^{m-1} (-)^i\,{}_{m-1}C_i\,g^{m-i}\,f^{(i)} + \sum_{i=0}^{m-1} (-)^{i+1}\,{}_{m-1}C_i\,g^{m-i-1}\,f^{(i+1)}. \qquad [110]$$

The <u>last</u> term of [110] can be written as:

$$\sum_{j=1}^{m-1} (-)^j\,{}_{m-1}C_{j-1}\,g^{m-j}\,f^{(j)} + (-)^m\,{}_mC_m\,f^{(m)}$$

where we have changed indices i into j-1 and used the fact that ${}_{m-1}C_{m-1} = {}_mC_m$ and $g^o = \hat{1}_V$. Therefore, after a change of indices [110] becomes

$$fg^m = \left\{ (-)^o\,{}_{m-1}C_o\,g^m\,f^{(o)} + \sum_{j=1}^{m-1} (-)^j\,{}_{m-1}C_j\,g^{m-j}\,f^{(j)} \right\} +$$

$$+ \left\{ \sum_{j=1}^{m-1} (-)^j\,{}_{m-1}C_{j-1}\,g^{m-j}\,f^{(j)} + (-)^m\,{}_mC_m\,f^{(m)} \right\}. \qquad [111]$$

Using the identities ${}_{m-1}C_o = {}_mC_o$ and ${}_mC_j = {}_{m-1}C_j + {}_{m-1}C_{j-1}$, we can write [111] into:

$$f\, g^m = \sum_{j=o}^{m} (-)^j {}_mC_j\, g^{m-j}\, f^{(j)} \qquad [112]$$

which completes the induction. ||

Proposition XVII

Let f and g be two Γ-endos on V such that

$$\underbrace{[g, \ldots [g, [g,}_{n\ \text{copies}} f]] \ldots] = \hat{0}_V , \qquad [113]$$

for some positive integer n, then the Fitting components of V, w.r.t. g, are f-invariant.

Proof

In the notation of Proposition XVI, [113] is simply

$$f^{(i)} = \hat{0}_V , \quad \text{for} \quad i \geq n. \qquad [114]$$

Step 1

We want to show that $V_{g,o}$ is f-invariant. Let m be the lowest upper bound of the integers i in the definition of

$$V_{g,o} = \{ z \mid z \in V,\ g^i z = 0 \ \text{ for some } \ i \in I_+ \}. \qquad [115]$$

From Proposition XVI we have,

$$g^{n+m}f = \sum_{i=0}^{n+m} {}_{n+m}C_i \, f^{(i)} \, g^{n+m-i} \qquad [116]$$

$$= \sum_{i=0}^{n-1} {}_{n+m}C_i \, f^{(i)} \, g^{n+m-i} \qquad \text{(use [113])} \qquad [117]$$

$$= \sum_{j=n-1}^{0} {}_{n+m}C_{n-j} f^{(n-j)} g^{m+j} \qquad (i \longrightarrow n-j) \qquad [118]$$

i.e. $$g^{n+m}f = \sum_{j=0}^{n-1} {}_{n+m}C_{n-j} f^{(n-j)} g^{m+j}. \qquad [119]$$

By [114], $g^{m+j}z = 0$ for any $z \in V_{g,o}$. Hence [119] yield

$$(g^{n+m} f) z = 0 \qquad [120]$$

i.e. $$g^{n+m}(fz) = 0 \qquad [121]$$

or $$fV_{g,o} \subset V_{g,o} . \qquad [122]$$

Consequently $V_{g,o}$ is f-invariant. ▮

Step 2

We want to show that $V_{g,1}$ is f-invariant. Let m be the

integer defined in Step 1. Let s be the lowest upper bound of the integers i such that

$$g^{i}V = g^{i+1}V = \cdots = V_{g,1} \quad . \tag{123}$$

Further, let r be the larger number of m and s. Consider $f\,g^{n+r}$, we have, by [105],

$$f\,g^{n+r} = \sum_{i=0}^{n+r} (-)^{i}\; {}_{n+r}C_{i}\; g^{n+r-i}\, f^{(i)} \tag{124}$$

$$= \sum_{j=0}^{n-1} (-)^{n-j}\; {}_{n+r}C_{n-j}\, g^{r+j} f^{(n-j)} \tag{125}$$

after some manipulation similar to that of [119]. By [123] we have

$$(g^{r+j}f^{(n-j)})V \subset g^{r+j}V = V_{g,1} \,. \tag{126}$$

Hence it follows that

$$(\text{RHS of [125]})\,V \subset V_{g,1} \tag{127}$$

i.e. $$(fg^{n+r})\,V \subset V_{g,1} \tag{128}$$

or $$fV_{g,1} \subset V_{g,1} \tag{129}$$

i.e. $V_{g,1}$ is f-invariant. ||

§. 5.5. Some properties of polynomials of a K-endo.

Let V denote a K-module throughout this section and the remainder of the chapter.

For any given $f \in End_K V$, we shall denote by $p[f]$ a polynomial of f with coefficients in K. The set

$$V_{p[f]} \equiv \{ v \mid v \in V,\ (p[f])^i v = 0, \quad \text{for some } i \in I_+ \} \qquad [130]$$

can be defined as the Fitting nil-component w.r.t. $p[f]$.

It is easy to show that $V_{p[f],o}$ is an f-invariant submodul of V. For any $x, x' \in V_{p[f],o}$ and $k, k' \in K$, we have

$$(p[f])^i (k \square x + k' \square x') = k \square ((p[f])^i x) + k' \square ((p[f])^i x')$$
$$= k \square 0 + k' \square 0 = 0 \quad . \qquad [131]$$

where s is any integer large enough such that

$$(p[f])^i V_{p[f],o} = \{0\}, \quad \text{for } i \geq s \quad . \qquad [132]$$

Therefore $V_{p[f],o}$ is a submodule of V (since other axioms are obviously satisfied by $V_{p[f],o}$). Further, it is f-invariant since, for any $x \in V_{p[f],o}$,

$$(p[f])^s (fx) = ((p[f])^s \circ f)x = (f \circ (p[f])^s)x = f(0) = 0. \qquad [13$$

Proposition XVII can be generalized to the case of a polynomial:

Proposition XVIII

Let f and g be K-endos on V such that

$$\underbrace{[g, \dots [g, [g,}_{i \text{ copies}} f]] \dots] = \hat{0}_V \;, \quad \text{for} \quad i \geq n \;, \qquad [134]$$

then, for any polynomial $p[g]$, the Fitting nil-component $V_{p[g],o}$ is f-invariant.

Proof

Prelude

First we need the following two facts (whose proofs are left as exercises to the reader. Or, see the Problem Section):

i) Let $K[\lambda]$ be the ring of polynomials of a single indeterminate λ, with coefficients in K. Let $\{\sigma_i\}_{i \in I_+}$ be a set of K-linear mappings defined by

$$\sigma_i : K[\lambda] \longrightarrow K[\lambda] \qquad [135]$$

with $\sigma_i : \lambda^j \longmapsto {}_jC_i\, \lambda^{j-i}$, for any $i, j (j \geq i) \in I_{(+)}$ [136]

where ${}_jC_i \equiv 0$ if $i > j$. For economy of notation we shall often write a polynomial $p[\lambda]$ as p. Then, for any $p, p' \in K[\lambda]$,

$$\sigma_j(pp') = \sum_{i=0}^{j} \sigma_i(p)\, \sigma_{j-i}(p') . \qquad [137]$$

ii) Let us write $p \| p'$ if p is a factor of p' (i.e. there exists a polynomial p'' such that $p' = pp''$). Then, for every p and $p' \in K[\lambda]$,

$$p^{j+1} \| p' \Longrightarrow p \| \sigma_i(p') , \quad i = 0, 1, \dots, j . \qquad [138]$$

Step 1

Let s be the integer defined by [132], i.e.

$$(p[g])^i V_{p[g],o} = \{0\} , \quad \text{for} \quad i \geq s . \qquad [139]$$

Consider now the polynomial $q \equiv p^{sn}$ where n is defined by [134 We now expand ($q = q[\lambda]$):

$$q \equiv \sum_{i=0}^{r} k_i \lambda^i , \qquad [140]$$

and define a corresponding polynomial in g:

$$q[g] \equiv \sum_{i=0}^{r} k_i g^i . \qquad [141]$$

The integer r is clearly finite since p is a polynomial. Analogous to the mapping [136], define σ_i to be the K-linear mapping

$$\sigma_i : K[g] \longrightarrow K[g] \qquad [142]$$

with $\sigma_i : g^j \longmapsto {}_jC_i g^{j-i}, \quad j(\geq i) \in I_+ ,$ [143]

where ${}_jC_i \equiv 0$ for $i > j$. By [141] and [106], we have

$$q[g]f = \sum_{i=0}^{r} \sum_{j=0}^{i} k_i \, {}_iC_j \, f^{(j)} g^{i-j} \qquad [144]$$

$$= \sum_{i=0}^{r} \sum_{j=0}^{r} k_i \, {}_iC_j \, f^{(j)} g^{i-j} , \qquad [145]$$

since ${}_iC_j = 0$ for $i < j$.

In terms of σ_i, [145] becomes

$$q\,[g]f = \sum_{j=0}^{r} f^{(j)} \; \sigma_j(q[g]) . \qquad [146]$$

But [134] is identical to the expression

$$f^{(i)} = 0 , \quad \text{for } i \geq n , \qquad [147]$$

hence [146] is effectively

$$q[g]f = \sum_{j=0}^{n-1} f^{(j)} \; \sigma_j(q[g]) . \qquad [148]$$

Step 2

Since $q \equiv p^{sn}$, it is trivial that

$$p^{sn} \parallel q \, . \qquad [14$$

Hence, by [138],

$$p^{s} \parallel \sigma_j(q) \, , \quad \text{for} \quad j = 0, 1, \ldots, n-1 \qquad [15$$

i.e. $\exists$ polynomial $w_j[\lambda]$: $\sigma_j(q) = w_j p^{s}$, $j = 0, 1, \ldots, n-1$. [15

Corresponding to [151] we have

$$\sigma_j(q[g]) = (w_j[g])(p[g])^{s} \, . \qquad [15$$

Substitution of [152] into [148] yields

$$q[g]f = \left\{ \sum_{j=0}^{n-1} f^{(j)} w_j(g) \right\} (p(g))^{s} \qquad [15$$

i.e. $(q[g]f)V_{p[g],o} = \{0\}$ (by [139])

or $$(p[g])^{sn}(fV_{p[g],o}) = \{0\} \, . \qquad [15$$

Hence,

$$f\,V_{p[q],o} \subset V_{p[q],o} \, . \qquad [1$$

This completes the proof. ||

§. 5.6. Minimal polynomial of a K-endo

Let V be a K-module of dimension n, then it is not difficult to show that

$$\dim(\mathrm{End}_K V) = n^2 . \qquad [156]$$

Hence the K-module $\mathrm{End}_K V$ is finite-dimensional if V is. [156] follows from the following proposition:

Proposition XIX

Let U and V be K-modules with dim.U = m and dim.V = n, then

$$\dim(\mathrm{Hom}_K(U,V)) = mn . \qquad [157]$$

Proof

Let $\{x_1, \ldots, x_m\}$ be a base of U, and $\{y_1, \ldots, y_n\}$ be a base of V. Introduce the K-homs f_{ij} defined by

$$f_{ij} : U \longrightarrow V$$

$$\text{with} \quad f_{ij} : x_k \longmapsto y_j\, \delta_{ki}, \quad i,k = 1,\ldots,m \text{ and } j = 1,\ldots,n. \qquad [158]$$

We want to show that the set $\{f_{ij}\}$ forms a base of $\mathrm{Hom}_K(U, V)$. For every $h \in \mathrm{Hom}_K(U, V)$, we can write hx_i in terms of y_j:

$$hx_i \equiv \sum_{j=1}^{n} a_{ij} y_j \quad \text{with} \quad a_{ij} \in K \, , \qquad [15$$

where $i = 1, \ldots, m$. We find, for every $k = 1, \ldots, m$, that

$$\left(\sum_{i=1}^{m} \sum_{j=1}^{n} a_{ij} f_{ij} \right) x_k = \sum_{i=1}^{m} \sum_{j=1}^{n} a_{ij} y_j \, \delta_{ki} = \sum_{j=1}^{n} a_{kj} y_j = hx_k \, .$$

hence

$$h = \sum_{i=1}^{m} \sum_{j=1}^{n} a_{ij} f_{ij} \qquad [16$$

i.e. $\{f_{ij}\}$ is a set of <u>generators</u> of $\mathrm{Hom}_K(U, V)$. It can b shown that the set is K-linearly independent. Suppose there i a set of coefficients $c_{ij} \in K$ such that

$$\sum_{i=1}^{m} \sum_{j=1}^{n} c_{ij} f_{ij} = \hat{0} \qquad [16$$

then, for any x_k,

$$\sum_{i=1}^{m} \sum_{j=1}^{n} c_{ij} f_{ij}(x_k) = 0$$

i.e.
$$\sum_{i=1}^{m} \sum_{j=1}^{n} c_{ij} y_j \, \delta_{ki} = \sum_{j=1}^{n} c_{kj} y_j = 0$$

i.e. $\quad c_{kj} = 0\ ,\quad$ for every k and j, [162]

since $\{y_j\}$ is a base. Consequently, we established that $\{f_{ij}\}$ is a base of $\mathrm{Hom}_K(U, V)$. Thus

$$\dim(\mathrm{Hom}_K(U, V)) = \mathrm{card}\left(\{f_{ij}\}_{\substack{i = 1,\ldots,m \\ j= 1,\ldots,n}}\right) = mn\ . \qquad [163]$$

∎

Let f be a K-endo on V. For convenience, we introduce the notation:

$$f^0 \equiv \hat{1}_V \quad \text{and} \quad f^1 \equiv f\ . \qquad [164]$$

Further, let s be an integer such that

$$\{f^0, f^1, \ldots, f^{s-1}\} \text{ is K-linearly \underline{independent}} \qquad [165]$$

but $\{f^0, f^1, \ldots, f^{s-1}, f^s\}$ is K-linearly dependent. [166]

Hence we can write f^s as

$$f^s = \sum_{i=1}^{s-1} a_i f_i\ , \quad a_i \in K\ . \qquad [167]$$

We note that the integer s is finite, by [156], since only finite dimensional K-modules are considered here.

Obviously, [167] can be restated in the language of "ring of polynomials" (cf. Chapter III). Let $K[t]$ denote the polynomial ring in an indeterminate t with coefficients in K, then $K[t]$ is also a K-module. To relate $K[t]$ to [167] we introduce the mapping

$$\mu : K[t] \longrightarrow \mathrm{End}_K V \qquad [168]$$

with $\mu : p[t] \longmapsto p[f]$, [169]

for every $p[t] \in K[t]$. We note that $\mu(t^o) \equiv \hat{1}_V$. It is easy to see that μ is a K-hom.

The particular polynomial, $\mathbb{P}_f(t)$, defined by

$$\mathbb{P}_f(t) \equiv t^s - \sum_{i=o}^{s-1} a_i\, t_i \qquad [170]$$

is called the minimal polynomial of f. (The a_i's are defined by [167]) It is obvious that

$$\mathbb{P}_f(f) = \hat{0}\ . \qquad [171]$$

Hence

$$\mu : \mathbb{P}_f(t) \longmapsto \mathbb{P}_f(f) = \hat{0}$$

i.e. $\mathbb{P}_f(t) \in \mathrm{Ker.}\,\mu$. [172]

(def) If a minimal polynomial is factorized into

$$\mathbb{P}_f(t) \equiv \prod_{i=1}^{m} (\rho_i(t))^{n_i}, \quad \text{with} \quad \rho_i(t) \neq \rho_j(t) \quad \text{if} \quad i \neq j, \qquad [173]$$

and such that each $\rho_i(t)$ cannot be further factorized, then each polynomial $\rho_i(t)$ is called an irreducible factor of the minimal polynomial of the K-endo f .

Let $f \in \mathrm{End}_K V$. Similar to [165] and [166] we now pick up any fixed non-zero $x \in V$ and construct the set

$$\{x, fx, f^2x, \ldots, f^m x\} \equiv x[f] \qquad [174]$$

such that $x[f]$ is K-linearly independent and

$$\{x, fx, f^2x, \ldots, f^m x, f^{m+1}x\}$$

is K-linearly dependent. Then we can define:

(def) For any given $f \in \mathrm{End}_K V$ and a non-zero $x \in V$, the sub K-module generated by $x[f]$ is called the cyclic K-module generated by x w.r.t. f . It is clear that $x[f]$ is a base of the cyclic K-module just mentioned. A K-module V is said to be cyclic w.r.t. f if there exists an $x \in V$ such that V is the cyclic K-module generated by x w.r.t. f .

By [174], we have

$$\exists a_i \in K : \sum_{i=1}^{m+1} a_i f^i x = 0$$

or $$\left(\sum_{i=1}^{m+1} a_i f^i\right) x = 0$$

Denote $\sum_{i=1}^{m+1} a_i f^i$ by $T_x(f)$ and $\sum_{i=1}^{m+1} a_i t^i$ by T_x(or $T_x(t)$).

Then one can show that

$$V \text{ is } \underline{\text{cyclic}} \iff \exists f \in \mathrm{End}_K V : \deg.\mathbb{P}_f(t) = \dim.V .$$

We leave the proof to the reader (or see the Problem Section).

§. 5.7. Eigen vectors and eigen values of a K-endo.

U and V denote K-modules in this section.

We remind the reader (cf. Chapter IV) that a K-module V is also called a vector space over K (or a K-space), and the elemen of V are often called vectors.

(def) Let f be a K-endo on V, then a non-zero $x \in V$ is calle an eigen vector of f with eigen value $\lambda \in K$ if

$$fx = \lambda x . \qquad [175]$$

[175] is called an eigen equation for f. The set of all eigen values of f is called the spectrum of f. In general, there ma be more than one eigen vector having the same eigen value, then we say the eigen value is degenerate. An eigen value has only

one corresponding eigen vector is said to be non-degenerate. For a given $f \in End_K V$, the set

$$V_\lambda = \{0_V\} \bigcup \{x | x \in V,\ fx = \lambda x\} \qquad [176]$$

is called the eigen space (i.e. eigen submodule, of V) w.r.t. f and λ .

It is not difficult to see that V_λ is an f-invariant submodule of V . Zero is the only element, in V_λ , that is not an eigen vector. It is joined into [176] in order to form a submodule.

(def) $\dim.V_\lambda \equiv$ geometric multiplicity of λ . [177]

(def) $f \in Hom_K(U, V)$ is said to be singular if it has no inverse, i.e.

$$\nexists f^{-1} \in Hom_K(V, U) : f^{-1} \circ f = \hat{1}_U . \qquad [178]$$

Otherwise we say that f is non-singular.

Remarks

i) For a singular f , the corresponding eigen space is just ker.f .

ii) $\hat{0}_V$ is nilpotent (trivial!) and has V as its eigen space.

Proposition XX

A K-hom f, from U to V, is non-singular iff f is a K-iso.

Proof

First, if f is non-singular then f^{-1} exists. Clearly, f must be bijective in order to produce f^{-1}. Hence f is a K-iso. ▌

Next, if f is a K-iso, then we reverse the mapping f to get f^{-1}. Clearly, f^{-1} is bijective and $f^{-1} \circ f = \hat{1}_U$. It is only necessary to show that f^{-1} is a K-hom. Consider any $\lambda_1, \lambda_2 \in K$ and $v_1, v_2 \in V$. Then the bijectivity of f allows us to write

$$v_1 = f(u_1) \text{ and } v_2 = f(u_2), \text{ for some } u_1, u_2 \in U . \qquad [179]$$

Thus

$$\begin{aligned} f^{-1}(\lambda_1 v_1 + \lambda_2 v_2) &= f^{-1}(\lambda_1 f(u_1) + \lambda_2 f(u_2)) \\ &= f^{-1}(f(\lambda_1 u_1) + f(\lambda_2 u_2)) \\ &= f^{-1}(f(\lambda_1 u_1 + \lambda_2 u_2)) \\ &= \lambda_1 u_1 + \lambda_2 u_2 = \lambda_1 f^{-1}(v_1) + \lambda_2 f^{-1}(v_2) . \end{aligned}$$ ‖

Proposition XXI

Let f be a K-endo on V, then

λ is an eigen value of $f \Longleftrightarrow f - \lambda\hat{1}_V$ is singular. [180]

<u>Proof</u>

<u>The necessity proof</u>

Let a non-zero $v_\lambda \in V$ be an eigen vector of λ , i.e.

$$fv_\lambda = \lambda v_\lambda .$$

Then $(f - \lambda\hat{1}_V)v_\lambda = 0$. [181]

That is, $\ker(f - \lambda\hat{1}_V)$ contains a <u>non-zero</u> element v_λ , hence $(f - \lambda\hat{1}_V)$ cannot be an iso. Consequently, by Proposition XX, $(f - \lambda\hat{1}_V)$ is singular. ∎

<u>The sufficiency proof</u>

Let $f - \lambda\hat{1}_V \equiv f_\lambda$ be singular. If λ is <u>not</u> an eigen value of f then

$f_\lambda v \neq 0$, for every <u>non-zero</u> $v \in V$ [182]

i.e. $\ker . f_\lambda = \{0\}$.

Hence f_λ is <u>injective</u>. But f_λ is a K-endo, injectivity implies immediately surjectivity. Therefore, f_λ is a K-iso. Consequently, by Proposition XX, f_λ is non-singular which contradicts our assumption. Hence λ must be an eigen value of f.

Remark

For $\lambda = 0$, [180] is reduced to the following special case:

If f is singular (resp. non-singular), then zero is (resp. is not) an eigen value of f.

Proposition XXII

If λ is an eigen value of $f \in \mathrm{End}_K V$, then its correspondi eigen space is $\ker(f - \lambda \hat{1}_V)$.

Proof

We leave this to the reader. ||

Proposition XXIII

Let λ be an eigen value of a K-endo f, on V, and let P(be any polynomial of f. Then $P(\lambda)$ is an eigen value of $P(f)$.

Proof

Let v be an eigen vector of f_λ, i.e. $fv = \lambda v$. Let $P(f)$ be written as

$$P(f) \equiv \sum_{i=0}^{m} a_i f^i \quad , \quad a_i \in K \quad \text{and} \quad f^0 \equiv \hat{1}_V \; . \qquad [183]$$

Then,

$$P(f)v = \sum_{i=0}^{m} a_i f^i v = \sum_{i=0}^{m} a_i \lambda^i v = P(\lambda)v \; . \quad || \qquad [184]$$

Remark

For $\lambda = 0$, the above theorem is reduced to the statement:

If f is nilpotent (i.e. $\exists n : f^n = \hat{0}_V$) then zero is the only eigen value of f.

Proposition XXIV

Let $f \varepsilon \mathrm{End}_K V$, then the set of all eigen vectors (of f) with distinct eigen values is K-linearly independent.

Proof

We prove the theorem by contradiction. Denote by $X = \{x_1, \ldots, x_r\}$ the set of all eigen vectors of f with corresponding distinct eigen values $\{\lambda_1, \ldots, \lambda_r\}$.
If we assume that X is K-linearly dependent, then we can choose a subset

$$X' = \{x_{i_1}, \ldots, x_{i_t}\} , \quad 2 \leq t \leq r , \qquad [185]$$

of X, such that any proper subset of X' is K-linearly independent. Since X is K-linearly dependent we have a unique non-trivial relation (i.e. not every a_j is zero below):

$$\sum_{j=1}^{t} a_j x_{i_j} = 0 , \quad \text{for some } a_j \varepsilon K \qquad [186]$$

$$\text{i.e.} \qquad f\left(\sum_{j=1}^{t} a_j x_{i_j}\right) = 0$$

$$\text{or} \qquad \sum_{j=1}^{t} a_j \lambda_{i_j} x_{i_j} = 0 \qquad [187]$$

which contradicts the uniqueness of [186] since λ_{i_j} are distinc[t] and since the sum of [186] contains at least two terms (in order to be non-trivial). ‖

§. 5.8. Characteristic polynomial and eigen-space decomposition of a K-endo.

In this section, we shall make statements with reference to an arbitrary base of a K-module V of a finite dimension n. Under a base, any K-endo f on V is an $n \times n$ matrix with entries in K. In particular, the identity endo $\hat{1}$ is just $\underset{\sim}{1}$, the $n \times n$ unit matrix. In what follows we shall denote the matrix form (in a given base) of f by $\underset{\sim}{f}$.

(def) The characteristic polynomial of a K-endo f (on V) is defined as

$$P_f(t) \equiv \det(\underset{\sim}{f} - t\underset{\sim}{1}) \qquad [188]$$

where t denotes an indeterminate. $P_f(t)$ is a polynomial of

t with coefficients in K. The corresponding equation

$$P_f(t) = 0 \qquad [189]$$

i.e. $$\det(\underset{\sim}{f} - t\underset{\sim}{1}) = 0 \qquad [190]$$

is called the characteristic equation of f. It is easy to see that the expression $\det(\underset{\sim}{f} - t\underset{\sim}{1})$ is base-independent (i.e. independent of any particular choice of the base for V). Suppose $\underset{\sim}{T}$ is a matrix that maps one base to another. Then $\underset{\sim}{T}^{-1}$ exists (why?), and the corresponding change of $(\underset{\sim}{f} - t\underset{\sim}{1})$ is

$$(\underset{\sim}{f} - t\underset{\sim}{1}) \longmapsto \underset{\sim}{T}^{-1}(\underset{\sim}{f} - t\underset{\sim}{1})\underset{\sim}{T}\,. \qquad [191]$$

Clearly, both sides of [191] have the same determinant. In other words, $P_f(t)$ is base-independent.

(def) If the characteristic polynomial $P_f(t)$ can be factorized into

$$P_f(t) = \prod_{i=1}^{r} q_i(t) \qquad [192]$$

such that each polynomial $q_i(t)$ cannot be further factorized, then every $q_i(t)$ is called an irreducible factor of $P_f(t)$.

Proposition XXV

Let f be a K-endo on V, then λ_i is an eigen value of

f iff λ is a root of $P_f(t)$, i.e. a solution of $P_f(t) = 0$. In other words, the roots of the characteristic polynomials of f are the only eigen values of f.

Proof

By Proposition XXI, λ_i is an eigen value of f iff $(f - \lambda_i \hat{1})$ is singular. In a base, this means $\det(\underset{\sim}{f} - \lambda_i \underset{\sim}{1}) = 0$. Hence the theorem follows. ||

(def) Algebraic multiplicity of an eigen value.

If K is an algebraically closed field then it follows easily from the definition that $P_f(t)$ can be factorized into the product of linear polynomials:

$$P_f(t) = \prod_{i=1}^{r} (\lambda_i - t)^{m_i}, \quad \lambda_i \in K \qquad [193]$$

where t denotes the indeterminate. By Proposition XXV, we can see that $\{\lambda_i\}_{i=1,\ldots,r}$ are just the set of all distinct eigen values of f. Each eigen value λ_i of [193] is said to have an algebraic multiplicity m_i.

Proposition XXVI

For every eigen value λ_i of $f \in \mathrm{End}_K V$,

$$\begin{pmatrix}\text{algebraic multiplicity} \\ \text{of } \lambda_i\end{pmatrix} \geq \begin{pmatrix}\text{geometric multiplicity} \\ \text{of } \lambda_i\end{pmatrix}.$$

Proof

Denote by $f|_i$ and $\hat{1}_i$ the restrictions of f and $\hat{1}$ to the submodule V_{λ_i}. Let n_i be the geometric multiplicity of λ_i, i.e. $n_i \equiv \dim. V_{\lambda_i}$. By choosing a base $\{b_1, \ldots, b_n\}$ of V such that the subset $\{b_1, \ldots, b_{n_i}\}$ spans V_{λ_i}, one can easily see that

$$\det(\underset{\sim}{f}|_i - t\underset{\sim}{1}_i) \parallel \det(\underset{\sim}{f} - t\underset{\sim}{1}), \qquad [194]$$

where $\alpha \parallel \beta$ denotes that α is a factor of β. Then

$$\det(\underset{\sim}{f}|_i - t\underset{\sim}{1}_i) = (\lambda_i - t)^{n_i} \qquad [195]$$

i.e. $$(\lambda_i - t)^{n_i} \parallel \det(\underset{\sim}{f} - t\underset{\sim}{1}). \qquad [196]$$

If m_i is the algebraic multiplicity of λ_i then, by definition, $(\lambda_i - t)^{m_i}$ is the maximal power of $(\lambda_i - t)$ that can be factored out from $\det(\underset{\sim}{f} - t\underset{\sim}{1})$. Hence $m_i \geq n_i$, by [196]. ||

For convenience, let us introduce the notation

$$f_i \equiv f - \lambda_i \hat{1}. \qquad [197]$$

Then, the Fitting nil-component of V (w.r.t. f_i) is

$$V_{f_i,o} \equiv \{z \mid z \in V, (f_i)^s z = 0, \text{ for some } s \in I_+\}. \qquad [198]$$

It is easy to see that $V_{f_i,o}$ is an f-invariant submodule of V (we leave the verification to the reader). If a particular eigen value λ_i has a unit (algebraic) multiplicity, then [198] is reduced to

$$\{z \mid z \in V,\ fz = \lambda_i z\} \equiv V_{\lambda_i} \qquad [199]$$

which is just the eigen space of the corresponding eigen value λ_i. However, $V_{f_i,o}$ is generally not an eigen space of V. It is clear that $V_{f_i,o}$ properly contains V_{λ_i} when they are not identical.

Proposition XXVII

Let $f \in End_K V$ where K is an algebraically closed field. If λ_i (i=1, ..., r) are distinct eigen values of f with algebraic multiplicities m_i, then we have the unique decomposition

$$V = \bigoplus_{i=1}^{r} V_{f_i,o} \qquad [200]$$

with $$\dim . V_{f_i,o} = m_i \ , \qquad [201]$$

where $$f_i \equiv f - \lambda_i \hat{1} \ . \qquad [202]$$

<u>Proof</u>

For economy of notation, let us denote the Fitting <u>nil</u> and <u>iso</u> components $V_{f_i,o}$ and $V_{f_i,1}$ by $V_{i,o}$ and $V_{i,1}$. By Proposition XIV, they are invariant under f_i, hence under $f_i + \lambda_i \hat{1}$. But $f = f_i + \lambda_i \hat{1}$, it follows that $V_{i,o}$ and $V_{i,1}$ are invariant under f, hence under $f - t\hat{1}$. Since $V = V_{i,o} \oplus V_{i,1}$ we can choose a base $B = \{b_1, \ldots, b_n\}$ such that the subset $\{b_1, \ldots, b_m\}$ spans $V_{i,o}$ and $\{b_{m+1}, \ldots, b_n\}$ spans $V_{i,1}$. From the fact that $V_{i,o}$ and $V_{i,1}$ are $(f - t\hat{1})$-invariant, we have (in base B):

$$(\underset{\sim}{f} - t\underset{\sim}{1})\,V = \begin{pmatrix} F_{i,o} & 0 \\ 0 & F_{i,1} \end{pmatrix} \begin{pmatrix} V_{i,o} \\ V_{i,1} \end{pmatrix} . \qquad [203]$$

Consequently

$$\det(\underset{\sim}{f} - t\underset{\sim}{1}) = \det(F_{i,o})\,\det(F_{i,1}) , \qquad [204]$$

where $F_{i,o}$ and $F_{i,1}$ are the matrix forms of the restriction of $(f - \lambda\hat{1})$ to $V_{i,o}$ and $V_{i,1}$ respectively (in base B). By Proposition XXI and (iii) of Proposition XIV, λ_i is the <u>only</u> eigen value $F_{i,o}$ <u>can</u> have, while λ_i <u>cannot</u> be an eigen value of $F_{i,1}$. Hence, all the multiplicity of λ_i must appear in $F_{i,o}$. In other words,

$$\dim . V_{i,o} \geq (\text{algebraic multiplicity of } \lambda_i) = m_i . \qquad [205]$$

Since a determinant is base-independent the above argument does not depend on the base we use.

The above argument can be carried out first w.r.t. λ_1. Then we decompose $V_{1,1}$ into Fitting components w.r.t. λ_2, etc. Consequently,

$$V \supset V_{1,o} \oplus V_{2,o} \oplus \cdots \oplus V_{r,o} . \qquad [206]$$

From [205] and [196], we have

$$\dim\left\{\bigoplus_{i=1}^{r} V_{i,o}\right\} \geq \sum_{i=1}^{r} m_i = \dim.V , \qquad [207]$$

since K is algebraically closed. Dimensionality requirement forces [207] to be an equality, i.e.

$$\dim.V = \sum_{i=1}^{r} \dim.V_{i,o} \qquad [208]$$

and, consequently, [206] yields

$$V = \bigoplus_{i=1}^{r} V_{i,o} . \qquad [209]$$

||

Remark

[209] now forces [205] into equalities:

$$\dim . V_{i,o} = m_i \ , \quad i = 1, \ldots, r \qquad [210]$$

for an algebraically closed K.

Proposition XXVIII

Let K be algebraically closed and $f \in \mathrm{End}_K V$ be nilpotent with nilpotency index r. Then, for any given $x \in V$ such that $f^{r-1}x \neq 0$,

$$\{x, fx, f^2x, \ldots, f^{r-1}x\} \qquad [211]$$

is K-linearly independent.

Proof

We prove this by contradiction. Assume that the set [211] is not K-linearly independent. Let $k_i \in K$ be a set of elements, not all zeros, such that

$$\sum_{i=0}^{r-1} k_i f^i x = 0, \qquad [212]$$

with $f^o \equiv \hat{1}_V$ understood. If j is the smallest index with $k_j \neq 0$, then [212] can be written as

$$\sum_{i=j}^{r-1} k_j^{-1} k_i f^i x = 0$$

i.e. $$f^{r-(j+1)}\left(\sum_{i=j}^{r-1} k_j^{-1}k_i f^i x\right) = 0$$

or $$\sum_{s=1}^{r-(j+1)} k_j^{-1}k_{j+s}f^{s+r-1}x = - f^{r-1}x \qquad [213]$$

where we have made a change of indices in the last step. Since f has nilpotency index r, the LHS of [213] vanishes. Hence, $f^{r-1}x = 0$. This contradicts our assumption that $f^{r-1}x \neq 0$. Therefore, [211] is K-linearly independent. ||

(def) A *flag* of V is defined as a descending chain of submodu V_i with

$$V \equiv V_o \supset V_1 \supset \cdots \supset V_n = \{0\} \qquad [214]$$

and $\text{codim}.V_i = i$, for $i = 1, 2, \ldots, n$. ($n = \dim.V$)

A flag is said to be f-*invariant* if each V_i is f-invariant.

Proposition XXIX

Let K be algebraically closed, then for every given $f \in \text{End}_K V$ there is an f-invariant flag of V.

Proof

Since K is algebraically closed, the characteristic polynomial of f has at least one root in K. Let the corresponding eigen value be λ_1. If x_1 is an eigen vector corresponding to λ_1, i.e.

$$fx_1 = \lambda_1 x_1 , \qquad [215]$$

then $x_1 \neq 0$ by definition. Construct the K-module generated by x_1, to be denoted by $((x_1))$. Next, we consider the quotient module

$$V/((x_1)) \equiv U_2 . \qquad [216]$$

In U_2, we pick up any eigen vector u of $f|_{U_2}$, i.e. an element $x_2 \in V$ such that

$$fx_2 = \lambda_2 x_2 \bmod ((x_1)) \qquad [217]$$

and $$x_2 \neq 0 \bmod ((x_1)) . \qquad [218]$$

From $((x_1, x_2))$ we construct the quotient module

$$V/((x_1, x_2)) \equiv U_3 . \qquad [219]$$

Now, we pick up an eigen vector of $f|_{U_3}$, i.e. an element $x_3 \in V$ such that

$$fx_3 = \lambda_3 x_3 \bmod ((x_1, x_2)) \qquad [220]$$

and $$x_3 \neq 0 \bmod ((x_1, x_2)) . \qquad [221]$$

This process can only be repeated for n times since we can show that the set $((x_1, x_2, \ldots, x_n))$ is K-linearly independent ($n = \dim.V$). To see this we assume the contrary, i.e. there is a set of coefficients $\{k_i\}$ in K such that

$$\sum_{i=1}^{n} k_i x_i = 0 . \qquad [222$$

Let k_j be the non-zero coefficient with the largest subscript, then we can write [222] into

$$x_j = -\sum_{i=1}^{s-1} k_j^{-1} k_i x_i$$

i.e. $\quad x_j = 0 \bmod ((x_1, \ldots, x_{j-1}))$

which contradicts the way we picked up x_j (see [218] and [221] Hence the set $((x_1, \ldots, x_n))$ is K-linearly independent; the set obviously spans V.

The required flag can be defined in the following way:

$$V \equiv V_0 = ((x_1, \ldots, x_n)) \qquad [223$$

$$V_1 = ((x_1, \ldots, x_{n-1}))$$

$$\vdots$$

$$V_{n-1} = ((x_1)) .$$

It is easy to see that each V_i is f-invariant (why?). ||

(def) A base of V is called a triangularizing base w.r.t. $f \in End_K V$ if f is a triangular matrix under this base. (i.e. the matrix elements above, or else below, the diagonal are all zeros)

Proposition XXX

Let K be algebraically closed, then for each $f \in End_K V$ there exists a triangularizing base w.r.t. f.

Proof

Use the base $((x_1, x_2, \ldots, x_n))$ of [223]. From [215], [217], [220] and so forth, we have

$$
\begin{aligned}
fx_1 &= \lambda_1 x_1 \\
fx_2 &= \lambda_2 x_2 \mod ((x_1)) \\
&\equiv k_{21}x_1 + \lambda_2 x_2 , \quad \text{say} \quad (k_{21} \in K) \\
fx_3 &= \lambda_3 x_3 \mod ((x_1, x_2)) \\
&\equiv k_{31}x_1 + k_{32}x_2 + \lambda_3 x_3 , \quad \text{say} \quad (k_{31}, k_{32} \in K) \\
&\ldots\ldots .
\end{aligned}
\qquad [224]
$$

Since any arbitrary $x \in V$ can be written as

$$x \equiv \sum_{i=1}^{n} a_i x_i \qquad [225]$$

we have

$$f(x) = \sum_{i=1}^{n} a_i f(x_i) \ . \qquad [226]$$

Therefore, by [224], the matrix form of [226] is:

$$\underset{\sim}{f} \begin{pmatrix} a_1 \\ a_2 \\ \cdot \\ \cdot \\ \cdot \\ \cdot \\ a_n \end{pmatrix} = \begin{pmatrix} \lambda_1 & 0 & 0 & 0 \cdots 0 \\ k_{21} & \lambda_2 & 0 & 0 \cdots 0 \\ k_{31} & k_{32} & \lambda_3 & 0 \cdots 0 \\ \cdots & \cdots & \cdots & \cdots \\ \cdots & \cdots & \cdots & 0 \\ k_{n1} & k_{n2} & \cdots & \lambda_n \end{pmatrix} \begin{pmatrix} a_1 \\ a_2 \\ \cdot \\ \cdot \\ \cdot \\ \cdot \\ a_n \end{pmatrix} \qquad [227]$$

i.e. $\underset{\sim}{f}$ is a triangular matrix under the base [225]. ||

Proposition XXXI

Let $\lambda_1, \ldots, \lambda_n$ be the diagonal elements of the matrix of $f \in End_K V$, under a triangularizing base (w.r.t. f). Then the λ_i's form the set of all eigen values of f.

Proof

We leave this to the reader (or see the Problem Section).

Proposition XXXII (Cayley-Hamilton theorem)

Every K-endo satisfies its own characteristic equation. In other words, for every $f \in End_K V$,

$$P_f(f) = \hat{0} \; . \qquad [228]$$

Proof

Step 1

First, let us consider V as a K[t]-module where t is an indeterminate as before. The module composition is defined by

$$\Omega_{\blacksquare} : K[t] \times V \longrightarrow V \qquad [229]$$

with $$\Omega_{\blacksquare} : (p(t), v) \longmapsto [p(f)]\, v \, , \qquad [230]$$

for every $p(t) \varepsilon K[t]$ and $v \varepsilon V$. It is obvious that [230] defines a K[t]-module since, for any $p(t)$, $q(t) \varepsilon K[t]$,

$$[p(t)q(t)] \blacksquare v = [p(f)q(f)]v = p(f)[(q(f))v]$$
$$= p(f)[q(t) \blacksquare v] = p(t) \blacksquare [q(t) \blacksquare v] \; ,$$

and the other axioms are obviously satisfied. For convenience, let us introduce a K-base $X \equiv \{x_1, \ldots, x_n\}$ of V. Under this base, every $v \varepsilon V$ is an $n \times 1$ column matrix (n=dim.V), to be denoted by $\underset{\sim}{v}$. $\underset{\sim}{f}$ is an $n \times n$ matrix. Then [230] becomes:

$$p(t) \blacksquare \underset{\sim}{v} = [p(\underset{\sim}{f})]\, \underset{\sim}{v} \; . \qquad [231]$$

In particular,

$$t \blacksquare \underset{\sim}{v} = \underset{\sim}{f}\, \underset{\sim}{v} . \qquad [232]$$

Denote $(\underset{\sim}{f} - t\underset{\sim}{1})$ by $\underset{\sim}{A}(t)$. We have

$$\underset{\sim}{A}(t) \cdot \underset{\sim}{v} = \underset{\sim}{0} , \qquad [233]$$

where $\underset{\sim}{0}$ is the $n \times 1$ column matrix with zero entries. Define the matrix $\underset{\sim}{A'}(t)$ by

$$(\underset{\sim}{A'}(t))_{ij} \equiv (-)^{i+j} \det(\underset{\sim}{A}(t))_{\downarrow ji} \qquad [234]$$

where the subscript , $\downarrow ji$, signifies that j-th row and i-th column are deleted from $\underset{\sim}{A}(t)$. Then, one can easily show (see the Problem Section) that

$$\underset{\sim}{A'}(t)\underset{\sim}{A}(t) = (\det.\underset{\sim}{A}(t))\underset{\sim}{1} . \qquad [235]$$

[233] and [235] lead to

$$(\det.A(t)) \cdot \underset{\sim}{v} = \underset{\sim}{0}$$

i.e. $$P_f(t) \cdot \underset{\sim}{v} = \underset{\sim}{0} \qquad [236]$$

or $$P_f(f)\, v = \hat{0}\, v , \quad \text{for any } v \in V.$$

Hence $P_f(f) = \hat{0}$. ||

Proposition XXXIII

$$P_f(t) \,\|\, P_f(t) , \quad \text{for every } f \in \mathrm{End}_K V . \qquad [237]$$

i.e., the minimal polynomial is always a factor of the characteristic polynomial, of a given K-endo.

Proof

We leave it to the reader. Hint: use Cayley-Hamilton theorem.

§. 5.9. Primary-component decomposition w.r.t. a K-endo.

The so-called primary-component decomposition has the nice property that a K-module can be decomposed into a direct sum with unique summands once a K-endo is given (though these summands can, in general, be further decomposed). Throughout this section, t denotes an indeterminate.

(def) Primary components w.r.t. a K-endo.

For any given $f \in \mathrm{End}_K V$, let us write the minimal polynomial in the form:

$$\mathbb{P}_f(t) = \prod_{i=1}^{r} (\rho_i(t))^{n_i}, \quad n_i \in I_+ \qquad [238]$$

where ρ_i's are distinct irreducible polynomials with unit leading coefficients. Then the sets $V_{f;i}$ defined by

$$V_{f;i} \equiv \{x \mid x \in V, \ (\rho_i(f))^{n_i} x = 0\}, \quad i = 1, 2, \ldots, r, \qquad [239]$$

are called the primary components of V w.r.t. f. We note that the n_i appeared in [239] are defined by [238].

Proposition XXXIV

Each primary component of V w.r.t. f is an f-invariant submodule of V, i.e.

$$f V_{f;i} \subset V_{f;i} , \quad i = 1, \ldots, r . \tag{240}$$

Proof

It is obvious to see that $V_{f;i}$ is a submodule of V. Next by definition [239], for every $x_i \varepsilon V_{f;i}$,

$$(\rho_i(f))^{n_i}(fx_i) = [(\rho_i (f))^{n_i} \circ f] x_i = [f \circ (\rho_i (f))^{n_i}] x_i$$

$$= f((\rho_i(f))^{n_i} x_i) = f(0) = 0 .$$

i.e. $$fx_i \varepsilon V_{f;i} . \tag{241}$$

Hence $V_{f;i}$ is f-invariant. ||

Proposition XXXV

For every K-endo f on V,

$$V = \bigoplus_{i=1}^{r} V_{f;i} . \tag{242}$$

where r is the total number of primary components of V w.r.t.

Proof

Step 1

Let the minimal polynomial be factorized into its irreducible factors, as in [238],

$$\mathbb{P}_f(t) = \prod_{i=1}^{r} (\rho_i)^{n_i} \qquad [243]$$

where we write ρ_i in place of $\rho_i(t)$, for encomy of notation. Then the polynomials σ_i, defined by

$$\sigma_i \equiv \mathbb{P}_f/(\rho_i)^{n_i} \ , \qquad [244]$$

are mutually prime, i.e.

$$\sigma_i \nparallel \sigma_j \ , \quad \text{for every} \quad i \neq j \ . \qquad [245]$$

This allows us to find a set of r polynomials $\{\sigma_i'\}_{i=1,\dots,r}$ such that

$$\sum_{i=1}^{r} \sigma_i \sigma_i' = 1 \ . \qquad [246]$$

Introduce the notation $q_i \equiv \sigma_i \sigma_i'$. Then

$$\sum_{i=1}^{r} q_i = 1 \ . \qquad [247]$$

Step 2

We want to show that each q_i is a projection mapping. By means of the K-hom, μ, introduced in [169], we have

$$\sum_{i=1}^{r} q_i(f) = \hat{1} \qquad [248]$$

corresponding to [247]. For $i \neq j$,

$$\sigma_i \sigma_j = \mathbb{P}_f^{\,2}(\rho_i)^{-n_i}(\rho_j)^{-n_j} = \mathbb{P}_f \prod_{\substack{k=1 \\ (k\neq i,j)}}^{r} (\rho_k)^{n_k} . \qquad [249]$$

Under the K-hom μ, [249] yields

$$\sigma_i(f)\circ\sigma_j(f) = \mathbb{P}_f(f) \prod_{\substack{k=1 \\ (k\neq i,j)}}^{r} (\rho_k)^{n_k} . \qquad [250]$$

But Cayley-Hamilton theorem says that $\mathbb{P}_f(f) = \hat{0}$. Hence [250] implies

$$\sigma_i(f) \circ \sigma_j(f) = \hat{0},$$

i.e. $$q_i(f) \circ q_j(f) = \hat{0}, \quad \text{for every } i \neq j. \qquad [251]$$

On the other hand, $q_j(f)$[248] gives :

$$(q_j(f))^2 = q_j(f) \qquad [252]$$

in virtue of [251]. Therefore, $\{q_j\}_{j=1,\dots,r}$ is a set of projection mappings. Besides, they have two useful properties: (i) they are mutually orthogonal, in the sense of [251], (ii) they form a complete set, in the sense of [248]. The set $\{q_j\}_j$ is often called a resolution of unity.

Step 3

By means of the set $\{q_i(f)\}_i$, we can write

$$V = \left(\sum_{i=1}^{r} q_i(f)\right) V = \bigoplus_{i=1}^{r} V_i \qquad [253]$$

where $V_i \equiv q_i(f)\,V$. We want to show that V_i is precisely the $V_{f;i}$ defined by [239]. This can be done by showing set-inclusions in both directions. First, for every $x \in V_i$, we have

$$\exists v \in V : x = q_i(f)\,v \qquad [254]$$

$$\text{i.e.}\quad (\rho_i(f))^{n_i} x = (\rho_i(f))^{n_i} q_i(f)\,v = [(\rho_i(f))^{n_i}\sigma_i(f)\sigma_i'(f)]\,v$$

$$= \mathbb{P}_f(f)\,\sigma_i'(f)\,v = 0\,, \qquad [255]$$

where we have used [244] and Cayley-Hamilton theorem in the last two steps. [255] establishes that

$$V_i \subset V_{f;i}\,. \qquad [256]$$

Next, for any $y_i \in V_{f;i}$, the decomposition [253] gives

$$y_i \equiv \sum_{j=1}^{r} x_j , \quad x_j \in V_j . \qquad [257$$

By the definition of $V_{f;i}$, we have

$$(\rho_i(f))^{n_i} y_i = 0$$

i.e. $$\sum_{j=1}^{r} (\rho_i(f))^{n_i} x_j = 0 . \qquad [258$$

But V_i is f-invariant since $q_i(f)$ is a projection mapping and f commutes with every $q_i(f)$, thus [258] implies that

$$(\rho_i(f))^{n_i} x_j = 0 , \quad \text{for every } j = 1, \ldots, r . \qquad [259$$

By [256] and the definition of $V_{f;j}$, it follows from [259]

$$x_j = 0 \quad \text{if} \quad j \neq i \quad (i \text{ is fixed}) \qquad [260$$

i.e. $$y_i = x_i \in V_i \qquad [261$$

in virtue of [257]. Hence we have established that $V_{f;i} \subset V_i$ Therefore

$$V_{f;i} \subset V_i . \qquad [262$$

Summarizing [253], [256] and [262], we conclude:

$$V = \bigoplus_{i=1}^{r} V_{f;i} \quad . \blacksquare$$

def) Let $f \in End_K V$, then the decomposition

$$V = \bigoplus_{i=1}^{r} V_{f;i} \qquad [263]$$

s called the primary-component decomposition of V w.r.t. f.

Now, it is natural to look for the relationship between the rimary-components and the minimal polynomial. The next theorem rovides this link.

roposition XXXVI

Denote by f_i the restriction of $f \in End_K V$ to the submodule $_{;i}$. Then the minimal polynomial of f_i is just the corresponding reducible factor of the minimal polynomial of f. In other words, $\mathbb{P}_f$ is factorized into irreducible factors,

$$\mathbb{P}_f = \prod_{i=1}^{r} (\rho_i)^{n_i} \ , \qquad [264]$$

en

$$\mathbb{P}_f = \prod_{i=1}^{r} \mathbb{P}_{f_i} \quad \text{and} \quad \mathbb{P}_{f_i} = (\rho_i)^{n_i} . \qquad [265]$$

Proof

For any $x \in V$, we can write

$$x = \sum_{i=1}^{r} x_i \;, \quad x_i \in V_{f;i} \qquad [266]$$

by the primary decomposition. From the fact that

$$\left\{ \prod_{i=1}^{r} \mathbb{P}_{f_i}(f) \right\} x_k = 0 \;, \quad \text{for } k = 1, \ldots, r \;,$$

it follows

$$\left[\prod_i \mathbb{P}_{f_i}(f) \right] x = 0$$

i.e.
$$\prod_i \mathbb{P}_{f_i}(f) = \hat{0} \;. \qquad [267]$$

However, since $\mathbb{P}_f(f)$ is the minimal polynomial of f, any function $A(t)$ such that $A(f) = \hat{0}$ implies that $\mathbb{P}_f \| A$. Hence [267] yields

$$\mathbb{P}_f \Big\| \prod_i \mathbb{P}_{f_i} \;. \qquad [268]$$

On the other hand, $\mathbb{P}_f$ is obviously the product of the minimal polynomials of f the restrictions of f, i.e.

$$\prod_i \mathbb{P}_{f_i} \parallel \mathbb{P}_f \, . \qquad [269]$$

[268] and [269] yield the equality

$$\mathbb{P}_f = \prod_{i=1}^{r} \mathbb{P}_{f_i} \, . \qquad [270]$$

It remains to show the second equality of [265]. As f_i is a restriction of f, therefore $\mathbb{P}_{f_i}$ must be of the form

$$\mathbb{P}_{f_i} = (\rho_i)^{m_i} \quad \text{with} \quad m_i \leq n_i \, , \qquad [271]$$

where n_i was defined as in [264]. From [264], [270] and [271] we conclude:

$$m_i = n_i \, , \quad i = 1, \ldots, r \, .$$

This establishes the second equality of [265]. ||

The next item we are interested is the relation between the Fitting decomposition and the primary decomposition, w.r.t. to a K-endo. The next theorem gives the needed information.

Proposition XXXVII

Let f be a K-endo on V. Let m be the non-negative integer such that $t^m \parallel P_f(t)$ and $t^{m+1} \nparallel P_f(t)$. For convenience , let us take the <u>first</u> primary component to be

$$\{x \mid x \in V,\ f^m x = 0\} \equiv V_{f;1} \,. \qquad [272]$$

Then the Fitting components are related to the primary components [263] by

$$V_{f,o} = V_{f;1} \quad \text{and} \quad V_{f,1} = \bigoplus_{i=2}^{r} V_{f;i}, \quad \text{for} \quad m \neq 0 \qquad [273]$$

and by

$$V_{f,o} = \{0\} \quad \text{and} \quad V_{f,1} = \bigoplus_{i=1}^{r} V_{f;i} \,, \quad \text{for} \quad m = 0 \,. \qquad [274]$$

<u>Proof</u>

If $m = 0$, then [272] is just the singleton $\{0\}$. Hence [274] follows from [80] and [97]. If $m \neq 0$, the situation is non-trivial. By [97] and [272] we see that

$$V_{f,o} \supset V_{f;1} \,. \qquad [275]$$

From the definition of $V_{f;1}$ we can also see that, for $i > 1$, t is <u>not</u> a factor of the minimal polynomial of f_i (i.e. the <u>restriction</u> of f to $V_{f;i}$). Hence f_i is non-singular, i.e. f_i is a K-iso. Thus

$$V_{f;i} \eqsim f_i V_{f;i} = f\, V_{f;i}$$

which yields, by iteration,

$$V_{f;i} = f\,V_{f;i} = f^2 V_{f;i} = \cdots \quad , \qquad [276]$$

for every $i > 1$. By [89], we have

$$V_{f,1} \supset V_{f;i} \quad , \quad \text{for every } \ i > 1 \ . \qquad [277]$$

Since $V_{f,1}$ is a K-module, it follows from [277]

$$V_{f,1} \supset \bigoplus_{i=2}^{r} V_{f;i} \ . \qquad [278]$$

On the other hand, we have

$$V_{f,o} \oplus V_{f,1} = \bigoplus_{i=1}^{r} V_{f;i} \ , \qquad [279]$$

therefore, in virtue of [275] and [278], we conclude:

$$V_{f,o} = V_{f;1} \quad \text{and} \quad V_{f,1} = \bigoplus_{i=2}^{r} V_{f;i} \ . \qquad \blacksquare$$

§. 5.10. Bilinear forms and inner products on K-modules

In this section, V and V' denote K-modules unless otherwise specified.

(def) A bilinear form f on $V \times V'$ is a mapping,

$$f : V \times V' \longrightarrow K, \text{ defined by} \qquad [280]$$

$$f : (x, x') \longmapsto f(x, x') ,$$

for every $x \in V$ and $x' \in V'$, such that

$$f(ax + by, cx' + dy')$$
$$= (ac)f(x,x') + (ad)f(x,y') + (bc)f(y,x') + (bd)f(y,y'), \qquad [281]$$

for $x,y \in V$; $x',y' \in V'$ and $a,b,c,d \in K$.

(notation) $Map_K(V \times V', K)$ = the set of all bilinear forms on $V \times V'$.

Proposition XXXVIII

$Map_K(V \times V', K)$ forms a K-module with a dimension equal to (dim.V)(dim.V').

Proof

It is obvious that $Map_K(V \times V', K)$ is a K-module w.r.t. th addition and the module-composition $\Omega_{\blacksquare}$ defined by:

$$(f + g)(x, x') = f(x,x') + g(x, x') \qquad [282$$

and $$(k \blacksquare f)(x, x') = k(f(x, x')) , \qquad [283$$

for every $f, g \in \mathrm{Map}_K(V \times V', K)$, $k \in K$, $x \in V$ and $x' \in V'$.

For the dimensionality proof, we take bases $B = \{b_1, \ldots, b_m\}$ and $B' = \{b'_1, \ldots, b'_n\}$, of V and V', respectively. Define a set of <u>bilinear forms</u> f_{ij} by

$$f_{ij}(b_r, b'_r) = \delta_{ir} \delta_{jt}, \qquad [284]$$

for $i, r = 1, \ldots, m$ and $j, t = 1, \ldots, n$. Then the set $\{f_{ij}\}_{i,j}$ forms a <u>base</u> of the K-module $\mathrm{Map}_K(V \times V', K)$. ▮

(def) A bilinear form f, on $V \times V$, is said to be <u>symmetric</u> if

$$f(x_1, x_2) = f(x_2, x_1) \qquad [285]$$

and <u>anti-symmetric</u> (i.e. <u>skew-symmetric</u>) if

$$f(x_1, x_2) = - f(x_2, x_1) \qquad [286]$$

for every $x_1, x_2 \in V$.

<u>Proposition XXXIX</u>

The set of all <u>symmetric</u> bilinear forms on $V \times V$ and the set of all <u>anti-symmetric</u> bilinear forms on $V \times V$ are submodules of the K-module $\mathrm{Map}_K(V \times V, K)$.

<u>Proof</u>

The proof is straightforward. We leave it to the reader.

(def) A bilinear form, f, on $V \times V'$ is said to be <u>non-degenera</u>
if, for any fixed non-zero $x \in V$,

$$\exists x' \in V' : f(x, x') \neq 0 \qquad [287]$$

and if, for any fixed non-zero $x' \in V'$,

$$\exists x \in V : f(x, x') \neq 0 . \qquad [288]$$

For the rest of the section, we shall restrict ourselves to the case where the ground field $\underset{\sim}{K}$ is $\mathbb{C}$ or $\mathbb{R}$, the complex-number field or the real-number field; W and W' shall denote $\underset{\sim}{K}$-modules .

(def) A mapping f on W is said to be <u>anti-linear</u> (or more precisely <u>anti-K-linear</u>) if, for every $x, y \in W$ and $a, b \in \underset{\sim}{K}$,

$$f(ax + by) = a^*f(x) + b^*f(y) \qquad [289]$$

where ( )* denotes the complex-number <u>conjugation</u>.

(def) A mapping f on $W \times W$ is said to be <u>hermitian</u> if f is <u>K-linear</u> w.r.t. its first argument and <u>anti-K-linear</u> w.r.t. its second argument, and such that

$$f^*(x, x') = f(x', x) , \qquad [290]$$

for every x and $x' \in W$.

The above definition amounts to :

$$f(ax + by,\ x') = af(x,\ x') + bf(y,\ x') \qquad [291]$$

and $$f(x',\ ax + by) = a^*f(x',\ x) + b^*f(x',\ y)\ . \qquad [292]$$

(def) A hermitian form f, on $W \times W$, is said to be positive-definite if, for every $x \in W$,

$$f(x,\ x) \geq 0 \qquad [293]$$

and $$x = 0 \Longleftrightarrow f(x,\ x) = 0\ . \qquad [294]$$

Similarly, negative-definite forms are defined by reversing the inequality sign of [293].

(def) A hermitian form, f, on $W \times W$ is said to be non-degenerate if, for any fixed non-zero $x \in W$,

$$\exists y \in W : f(x,\ y) \neq 0 \qquad [295]$$

and if, for any fixed non-zero $y \in W$,

$$\exists x \in W : f(x,\ y) \neq 0\ . \qquad [296]$$

Now we define the important concept of "inner product".

(def) An inner product, f, on W is a positive-definite

hermitian form on $W \times W$. $f(x, y)$ is called the inner product of x and y (w.r.t. f).

For convenience, we shall use the notation

$$\langle x, y\rangle \equiv f(x, y) , \qquad [297]$$

for the rest of this section.

(def) A K-module ($K = \mathbb{C}$ or $\mathbb{R}$) with an inner product is called an inner-product space.

With the introduction of "inner product" we can use a more geometrical language, for example:

(def) Two elements x and y of an inner-product space W are said to be orthogonal (w.r.t. the given inner product) if their inner product vanishes, i.e. $\langle x, y\rangle = 0$. An element $x \in W$ is said to be normalized if $\langle x, x\rangle = 1$. A subset U of W is orthonormal if each element of U is normalized and if any two elements of U are orthogonal.

(notation) We shall write $x \perp x'$ if x is orthogonal to x'.

For any subspace U of an inner-product space W, the set

$$U^{\perp} \equiv \{y \mid y \in W, \langle x, y\rangle = 0 \text{ for every } x \in U\} \qquad [298]$$

is a subspace (of W) to be called the orthogonal supplement of U (w.r.t. the given inner-product $\langle\, ,\, \rangle$).

Given any base $X \equiv \{x_1, \ldots, x_n\}$ of a $\mathbb{C}$-module W, one can use the Schmidt orthogonalization procedure to construct an orthonormal base of W. The procedure consists of first choosing any element of X, say $x_1 \varepsilon X$, to define

$$y_1 = x_1 \langle x_1, x_1 \rangle^{-1/2}$$

$$y_1' = x_2 - \langle x_2, y_1 \rangle y_1$$

$$y_2 = y_1' \langle y_1', y_1' \rangle^{-1/2}$$

$$\cdots\cdots\cdots\cdots$$

$$y_n = y_{n-1}' \langle y_{n-1}', y_{n-1}' \rangle^{-1/2}, \qquad [299]$$

then the set $\{y_1, \ldots, y_n\}$ is an orthonormal base of W. We leave the verification to the reader (or see the Problem Section).

Proposition XXXX

Let U be a subspace of an inner-product space W, then

$$W = U \oplus U^{\perp} \qquad [300]$$

and $$(U^{\perp})^{\perp} = U \qquad [301]$$

where $U^{\perp}$ is the orthogonal supplement of U.

Proof

Choose an orthonormal base $\{y_1, \ldots, y_m\}$ of U. Then, we can write every $x \in W$ into the form

$$x = \sum_{i=1}^{m} \langle x, y_i \rangle y_i + (x - \sum_{i=1}^{m} \langle x, y_i \rangle y_i) \qquad [302]$$

We can show that (i) the first sum in [302] is an element in U, (ii) the bracket in [302] is an element in $U^{\perp}$ and (iii) $U \cap U^{\perp} = \{0\}$. (i) is obvious since it is a K-linear combination of elements of U. (ii) follows the same line as in [P.18] of Problem 10. As to (iii), we have

$$\langle x, x \rangle = 0 ,$$

for every $x \in U \cap U^{\perp}$. Hence $x = 0$. ∎

Next, we want to show [301]. We have, by using [300] twic

$$\dim(U^{\perp})^{\perp} = \dim.W - \dim.U^{\perp}$$
$$= \dim.W - (\dim.W - \dim.U) = \dim.U .$$

This implies that $(U^{\perp})^{\perp}$ is K-isomorphic to U. But it is easy to see that $(U^{\perp})^{\perp} \supset U$, hence [301] follows. ∥

(def) A norm ω on W is a mapping,

$$\omega : W \longrightarrow \mathbb{R} \text{ , defined by}$$

$$\omega : x \longmapsto \|x\| \text{ ,} \qquad [303]$$

for every $x \in W$, such that

i) $\|kx\| = |k| \cdot \|x\|$, [304]

ii) $\|x\| + \|x'\| \geq \|x + x'\|$ [305]

iii) $\|x\| \geq 0$, and $x = 0 \Longleftrightarrow \|x\| = 0$, [306]

for every $x, x' \in W$ and $k \in \underset{\sim}{K}$.

It is clear that a norm can always be introduced in an inner-product space W by choosing

$$\|x\| \equiv \sqrt{\langle x, x \rangle} \text{ , for every } x \in W. \qquad [307]$$

It is easy to verify that [307] satisfies the axioms of a norm. It is important to note that the existence of a norm does not imply the existence of an inner product.

As another example, let us consider a base $\{b_1, \ldots, b_n\}$ of W. For every $x \in W$, we can write

$$x = \sum_{i=1}^{n} k_i b_i , \qquad k_i \in \underset{\sim}{K} .$$

Then we can define a norm by

$$\| x \| \equiv \max_i |k_i| . \qquad [308]$$

(def) A $\underset{\sim}{K}$-module equipped with a specific norm is called a *normed space.*

(notation) $\widetilde{V} \equiv \mathrm{Map}_{\underset{\sim}{K}}(V, \underset{\sim}{K})$. [309]

Now let us look into the property of $\widetilde{V} \equiv \mathrm{Map}_{\underset{\sim}{K}}(V, \underset{\sim}{K})$. If we define an addition and a module composition $\Omega_{\blacksquare}$ by

$$(f + f')x = f(x) + f'(x) \qquad [310]$$

and $\qquad (k \blacksquare f)x = k\,f(x)$, [311]

for every $f, f' \in \widetilde{V}$ and $x \in V$. Then, it is routine to verify that $\widetilde{V}$ is a $\underset{\sim}{K}$-module, w.r.t. $\Omega_{\blacksquare}$, to be called the *dual module* of V. We leave it to the reader to verify that

$$\dim.\widetilde{V} = \dim.V \qquad [312]$$

where we assumed V to be finite-dimensional. It can be shown (cf. Jacobson: Lectures in Abstract Algebra, vol. II, Chapter IX, Theorem 2), by set-theoretical arguments, that if V is an

<u>infinite-dimensional</u> Γ-module (where Γ is a division ring) then

$$\dim.\widetilde{V} = (\text{card.}\Gamma)^{\dim.V} \quad . \qquad [313]$$

[313] shows clearly

$$\dim.\widetilde{V} > \dim.V , \qquad [314]$$

for an infinite-dimensional V.

The "duality" concept can be extended to the bases of a $\underset{\sim}{K}$-module:

(def) Let V be a $\underset{\sim}{K}$-module and $\widetilde{V}$ be its dual. If

$$B \equiv \{b_1, \ldots, b_n\} \qquad [315]$$

is a base of V, then the set

$$\widetilde{B} = \{\widetilde{b}_i \mid \widetilde{b}_i \in \widetilde{V}, \widetilde{b}_i(b_j) = \delta_{ij}, \quad i, j, = 1, \ldots, n\} \qquad [316]$$

is called a base (of $\widetilde{V}$) dual to B. We have to verify that [316] is indeed a base of $\widetilde{V}$.

First, for any $f \in \widetilde{V}$, we have

$$f(b_i) \equiv k_i \in \underset{\sim}{K} \, . \qquad [317]$$

Hence

$$f(b_i) = \sum_{j=1}^{n} k_i \delta_{ij} = \sum_{j=1}^{n} k_j \tilde{b}_j(b_i) = (\sum_{j=1}^{n} k_j \tilde{b}_j) b_i$$

i.e. any $f \varepsilon \tilde{V}$ can be expressed as a $\underset{\sim}{K}$-linear combination of elements in $\tilde{B}$. It remains to show that $\tilde{B}$ is $\underset{\sim}{K}$-linearly independent. Let us assume

$$\sum_{i=1}^{n} c_i \tilde{b}_i = \tilde{0} , \quad c_i \varepsilon K , \qquad [318]$$

for some $c_i \neq 0$. $\tilde{0}$ denotes the zero-mapping on $\tilde{V}$. Then [318] yields

$$(\sum_i c_i \tilde{b}_i) b_j = \tilde{0}(b_j) = 0 , \quad j = 1, \ldots, n .$$

i.e. $\sum_i c_i \delta_{ij} = 0$ (by [316])

or $c_j = 0$, for $j = 1, \ldots, n$. [319]

Therefore $\tilde{B}$ is $\underset{\sim}{K}$-linearly independent. Hence $\tilde{B}$ is a base of $\tilde{V}$. In fact, this also shows

$$\dim.\tilde{V} = n = \dim.V \qquad [320]$$

for a finite-dimensional $\underset{\sim}{K}$-module V. ∎

The following proposition shows that there is a base-independent bijection between V and $\tilde{V}$:

<u>Proposition XXXXI</u>

Let $\tilde{V}$ be the dual of an inner-product $\underset{\sim}{K}$-space V, then

i) for every fixed $y \in \tilde{V}$,

$$\exists_{1!} x \in V : y(x') = \langle x', x \rangle, \quad \text{for every } x' \in V \qquad [321]$$

ii) for every fixed $x \in V$,

$$\exists_{1!} y \in \tilde{V} : y(x') = \langle x', x \rangle, \quad \text{for every } x' \in V. \qquad [322]$$

<u>Proof</u>

We give here only a proof of (i) since the proof of (ii) is similar. Let $\tilde{B} \equiv \{\tilde{b}_1, \ldots, \tilde{b}_n\}$ be the dual base of an <u>orthonormal</u> base $B \equiv \{b_1, \ldots, b_n\}$ of V.

$$\text{Let} \quad x \equiv \sum_i k_i b_i, \quad x' \equiv \sum_i k_i' b_i$$

$$\text{and} \quad y \equiv \sum_i c_i \tilde{b}_i. \qquad [323]$$

Then

$$y(x') = \sum_{i,j} c_i k_j' \tilde{b}_i(b_j) = \sum_{i,j} c_i k_j' \delta_{ij} = \sum_i c_i k_i' . \qquad [324]$$

On the other hand, we have

$$\langle x', x \rangle = \left\langle \sum_i k_i' b_i , \sum_j k_j b_j \right\rangle = \sum_{i,j} k_i' k_j^* \langle b_i, b_j \rangle$$

$$= \sum_{i,j} k_i' k_j^* \delta_{ij} = \sum_i k_i' k_i^* . \qquad [325]$$

If we choose

$$k_i' \equiv c_i^* , \qquad [326]$$

then [324] and [325] yield

$$y(x') = \sum_i c_i k_i' = \langle x', x \rangle \qquad [327]$$

which is just [321]. In other words the <u>unique</u> x is given by

$$x = \sum_{i=1}^{n} c_i^* b_i . \qquad \|$$ [328]

<u>Remark</u>

Though we have made use of bases in proving the above

proposition yet the bijection $\tilde{V} \longleftrightarrow V$ established by [326] is obviously base-independent since the inner-product appeared in [321] is independent of the choice of B.

§. 5.11. Group representations.

In this section G denotes a group and V a $\underset{\sim}{K}$-module. Hom refers to "group hom".

Because of the scope of the book, we shall give here a very brief account of some elementary notions on group representations, or equivalently, modules over a group.

(def) Any $\gamma \in \text{Hom}(G, \text{Aut}_{\underset{\sim}{K}} V)$ is called a representation of the group G. V is called the representation space or the G-module corresponding to γ. The equivalence of the notion of "representation" and "G-module" is now to be justified. The $\underset{\sim}{K}$-module V can be made into a G-module by means of the group-hom in the following manner. The required G-module composition,

$$\Omega_{\blacksquare} : G \times V \longrightarrow V \qquad [329]$$

is defined by

$$\Omega_{\blacksquare} : (g, x) \longmapsto g \blacksquare x \equiv (g^{\gamma})x , \qquad [330]$$

for every $g \in G$ and $x \in V$. g^{γ} denotes $\gamma(g)$, i.e.

$$\gamma : g \longmapsto g^{\gamma} \in \mathrm{Aut}_K V . \qquad [331]$$

It is straightforward to verify that [330] makes V a unitary G-module. (See the Problem Section)

(def) Let V be a G-module for a group G, then a sub K-module V' (of V) is called a sub G-module (or γ-invariant submodule where γ is the representation corresponding to the G-module V of V if

$$G \cdot V' \underset{\text{set}}{\subset} V' \qquad [332]$$

i.e. $g \cdot x' \in V'$, for every $g \in G$ and $x' \in V'$. [333]

Remark

We note that [332] is equivalent to

$$(\gamma(G))V' \underset{\text{set}}{\subset} V' \qquad [334]$$

i.e. $(g^{\gamma})\, x' \in V'$, for every $g \in G$ and $x' \in V'$.

(def) A representation $\gamma \in \mathrm{Hom}(G, \mathrm{Aut}_K V)$ is said to be irreducible if V, as the corresponding G-module, is simple; that is, V is non-zero and contains no non-trivial sub G-modules. (Otherwise we say that γ is reducible)

(def) A representation $\gamma \in \mathrm{Hom}(G, \mathrm{Aut}_K V)$ is completely reducible if V, as the corresponding G-module, is semi-simple; that is, if every sub G-modules (of V) is a direct summand with a supplementary sub G-module. In other words, for every sub G-module V_1, there exists a sub G-module V_2 such that

$$V = V_1 \oplus V_2 . \qquad [335]$$

Because of the conceptual equivalence of representations and G-modules, we have the following identification of languages:

i) an irreducible representation $\Longleftrightarrow$ the corresponding G-module is simple.

ii) a completely reducible representation $\Longleftrightarrow$ the corresponding G-module is semi-simple.

Proposition XXXXII

Let V be a G-module with composition $\Omega_{\blacksquare}$. If there exists a projection mapping p of V onto a sub K-module U of V such that

$$p(g \blacksquare x) = g \blacksquare (px), \quad \text{for every } g \in G, \qquad [336]$$

then both U and $(\hat{1}_V - p)V$ are sub G-modules of V.

Proof

First, we want to show that U is a sub G-module. Since p is a projection mapping we have, for any $y \in U$,

$$\exists\, x \in V : y = p(x) .$$

i.e. $g \blacksquare y = g \blacksquare (px)$, for every $g \in G$

or $g \blacksquare y = p(g \blacksquare x) \in pV = U$.

Hence U is a sub G-module. ∎

Next, we want to show that $(\hat{1}_V - p)\,V$ is also a sub
G-module. It is obvious that, for any $z \in (\hat{1}_V - p)\,V$,

$$\exists\, x \in V : z = x - px$$

i.e. $g \blacksquare z = g \blacksquare (x - px)$, for every $g \in G$

or $g \blacksquare z = g \blacksquare x - g \blacksquare (px) = g \blacksquare x - p(g \blacksquare x)$

$$= (\hat{1}_V - p)(g \blacksquare x) \in (\hat{1}_V - p)\,V .$$

Thus $(\hat{1}_V - p)\,V$ is a sub G-module. ∎∎

Problems with hints or solutions for Chapter V

Problem 1

Carry out the sufficiency proof of Proposition III.

Proof

Let K-modules V and W be of the same dimension. If B and E are bases of V and W, respectively, then

$$\text{card}.B = \text{card}.E \quad .$$

This allows us to introduce a bijection f from B to E. Since f can be extended to V by K-linearity, the extension establishes a K-iso between V and W. ||

Problem 2

Let R be a ring. If there is an element $t \in R$ such that $rtr = r$ for every $r \in R$, show that tr' generates Rr' for any given $r' \in R$.

Proof

Obviously,

$$R(tr') \subset Rr' \qquad [P.1]$$

since $Rt \subset R$. Next, from $r\,r' \in R$, we have

$$(rr')tr' \in Rtr' .$$

Hence

$$rr' = r(r'tr') = (rr')tr' \in Rtr'$$

i.e. $\quad Rr' \subset R(tr')$. [P.2

[P.1] and [P.2] completed the proof. ||

Problem 3

Prove (1) and (2) of Proposition VI.

Proof

Proof of (1). It is obvious since $\dim.U \leq \dim.V$, and Proposi III rules out the equality sign. ▮

Proof of (2). We remind the reader that $U_1 \cap U_2$ is again submodule of U. Let B_1, B_2 and B_{12} be bases of U_1, U_2 an U_{12}. Then, by Proposition V,

$$\exists B_1' \subset B_1 : B_{12} \cup B_1' \text{ is a } \underline{\text{base}} \text{ of } U_1 , \qquad [P.3$$

$$\exists B_2' \subset B_2 : B_{12} \cup B_2' \text{ is a } \underline{\text{base}} \text{ of } U_2 . \qquad [P.4$$

We want to show that the set

$$B_{12} \cup B_1' \cup B_2' \equiv X \qquad [P.5$$

is a base of $U_1 + U_2$. Since X obviously generates $U_1 + U_2$ (by [P.3] and [P.4]), it is sufficient to show that X is Γ- linearly independent. Assume the contrary. Then there exists a particular linear combination, β, of B_2' such that β is expressible as a linear combination of $B_{12} \cup B_1'$. Since $B_{12} \cup B_1'$ is a base of U_1, β is expressible in terms of B_1. Hence $\beta \in U_1$. But $\beta \in U_2$ as implied by its definition, therefore $\beta \in (U_1 \cup U_2)$. In other words, β is a linear combination of B_{12}, therefore $B_{12} \cup B_2'$ is not Γ-linearly independent. This is impossible since $B_{12} \cup B_2'$ is a base (of U_2). Consequently, X is Γ-linearly independent, hence a base of $U_1 + U_2$. This leads to:

$$\begin{aligned} \dim(U_1 + U_2) &= \text{card}.X = \text{card}(B_{12} \cup B_1' \cup B_2') \\ &= \text{card}.B_{12} + \text{card}.B_1' + \text{card}.B_2' \\ &= \text{card}.B_{12} + (\text{card}.B_1 - \text{card}.B_{12}) + (\text{card}.B_2 - \text{card}.B_{12}) \\ &= \text{card}.B_1 + \text{card}.B_2 - \text{card}.B_{12} \\ &= \dim.U_1 + \dim.U_2 - \dim(U_1 \cap U_2) . \quad \| \qquad [P.6] \end{aligned}$$

Problem 4

Show [39] of Proposition VIII.

Proof

Since U and V_1 are both F-invariant, the intersection

$U \cap V_1$ is obviously F-invariant. $U \cap V_1$, as an F-invariant submodule of V_1 which is F-simple, must be either $\{0\}$ or V_1 itself. But [38] rules out the second possibility, hence $U \cap V_1 = \{0\}$. ||

Problem 5

Give the "sufficiency" proof of Proposition X.

Proof

Let π_i $(i = 1, \ldots, m)$ be projection operators defined by

$$\pi_i \circ \pi_j = \delta_{ij} \hat{1}_V \quad \text{and} \quad \sum_{i=1}^{m} \pi_i = \hat{1}_V \tag{P.7}$$

and such that $\pi_i \circ f = f \circ \pi_i$ for every $f \in F$. Then, by [P.7], we have

$$V = \sum_{i=1}^{m} \pi_i V = \bigoplus_{i=1}^{m} (\pi_i V) . \tag{P.8}$$

It is obvious that each $\pi_i V$ is F-invariant since

$$f(\pi_i V) = (f \circ \pi_i) V = (\pi_i \circ f) V = \pi_i (fV) \subset \pi_i V . \quad ||$$

Problem 6

Prove [137] of Proposition XVIII.

Proof

Consider any $p, q \in K[\lambda]$. Let their expansions be

$$p \equiv \sum_{i=0}^{\infty} a_i \lambda^i , \qquad [P.9]$$

$$p' \equiv \sum_{i=0}^{\infty} b_i \lambda^i \qquad [P.10]$$

with $a_i, b_i \in K$ and $a_i, b_i = 0$ p.p. $i \in I_{(+)}$. Then

$$\sum_{i=0}^{j} \sigma_i(p)\sigma_{j-1}(p') = \sum_{i=0}^{j} \sum_{r,s=0}^{\infty} a_r b_s \sigma_i(\lambda^r)\sigma_{j-i}(\lambda^s)$$

$$= \sum_{i=0}^{\infty} \sum_{r,s=0}^{\infty} a_r b_s ({}_rC_i \lambda^{r-i})({}_sC_{j-i} \lambda^{s+i-j})$$

$$= \sum_{r,s} a_r b_s (\sum_i {}_rC_i \, {}_sC_{j-i}) \lambda^{s+r-j}$$

$$= \sum_{r,s} a_r b_s \sigma_j(\lambda^{s+r}) = \sigma(pp')$$

which completed the proof. In the last second step we have used the identity (cf. L. B. Jolley, Summation of Series, Dover Publications, Inc, 1961, pp.36):

$$\sum_i {}_rC_i \, {}_sC_{j-i} = {}_{s+r}C_j \, . \qquad \|$$

Problem 7

Prove [138] of Proposition XVIII.

Proof

Use mathematival induction on the index j of [138]. The validity of [138] is clear for $j = 0$ since it gives the trivial relation

$$p \,\|\, p' \Longrightarrow p \,\|\, p' \qquad [P.11]$$

in virtue of the fact $\sigma_o(p') = p'$. By the induction hypothesis we now assume that [138] is true for $j = m - 1$, i.e.

$$p^m \,\|\, p' \Longrightarrow p \,\|\, \sigma_i(p') \ , \quad i = 0,1,\ldots,m \ . \qquad [P.12]$$

From the LHS of [138], i.e. $p^{m+1} \,\|\, p'$, we can write

$$p' \equiv pq \quad \text{with} \quad p^m \,\|\, q \ . \qquad [P.13]$$

By [P.13] and the induction assumption (cf. [P.12]), we have

$$p^m \,\|\, q \Longrightarrow p \,\|\, \sigma_i(q) \ , \quad i = 0,\ 1,\ \ldots,\ m \ . \qquad [P.14]$$

Hence it is trivial that

$$p \,\|\, \sigma_i(q)\, \sigma_{m-i}(p)$$

or

$$p \,\|\, \sum_{i=0}^{m-1} \sigma_i(q)\, \sigma_{m-i}(p)$$

i.e. $$p \,\Big\|\, \left\{ \sum_{i=0}^{m-1} \sigma_i(q)\, \sigma_{m-i}(p) + p \right\}$$

or $$p \,\Big\|\, \sum_{i=0}^{m} \sigma_i(q)\, \sigma_{m-i}(p) \,. \quad (p = \sigma_o(p)) \qquad [P.15]$$

Consequently, by [137],

$$p \,\|\, \sigma_m(qp) \qquad [P.16]$$

i.e. $$p \,\|\, \sigma_m(p') \qquad \text{(by [13])}$$

which completes the induction. ||

Problem 8

Show that (cf. the end of §.5.6)

$$V \text{ is } \underline{\text{cyclic}} \Longleftrightarrow \exists f \in \mathrm{End}_K V : \deg.\mathbb{P}_f(t) = \dim.V \,. \qquad [P.17]$$

Sketch of the "necessity" proof: Since [174] is K-linearly independent but the set $\{x, fx, \ldots, f^{m+1}x\}$ is not, we have

$$\exists a_i \in K : \sum_{i=1}^{m+1} a_i f^i x = 0 \qquad [P.18]$$

where at least one of the a_i is non-zero. Write

$$\alpha_x(f) \equiv \sum_{i=1}^{m+1} a_i f^i \ . \qquad [P.19]$$

Then

$$(\alpha_x(f))\, x = 0 \ . \qquad [P.20]$$

Now, if V is <u>cyclic</u>, then

$$\exists z \ \varepsilon\ V : V \equiv z[f] \qquad [P.21]$$

by the notation of [174]. It is easy to see that $\mathbb{P}_f(t) = \alpha_x(t)$. Hence

$$\deg.\mathbb{P}_f(t) = \deg\{\alpha_x(t)\} = \dim.V \ . \ \blacksquare \qquad [P.22]$$

<u>Sketch of the "sufficiency" proof</u>: One first shows that

$$\exists y \ \varepsilon\ V : \alpha_y(t) = \mathbb{P}_f(t) \ , \qquad [P.23]$$

then it follows that

$$\dim\{y[f]\} = \deg\{\alpha_y(t)\} = \deg.\mathbb{P}_f(t) = \dim.V \ . \qquad [P.24]$$

Therefore $y[f]$ must be identical to V since the former is a subset of the latter. $\blacksquare\blacksquare$

Problem 9

Prove Proposition XXXI of §.5.8.

Proof

$$P_f(t) = \det.(\underset{\sim}{f} - t\underset{\sim}{1})$$

$$= \det.\left\{\begin{pmatrix} \lambda_1 & & & 0 \\ & \lambda_2 & & \\ & & \ddots & \\ & & & \lambda_n \end{pmatrix} - \begin{pmatrix} t & & & 0 \\ & t & & \\ & & \ddots & \\ 0 & & & t \end{pmatrix}\right\}$$

$$= \begin{vmatrix} \lambda_1 - t & & & 0 \\ & \lambda_2 - t & & \\ & & \ddots & \\ & & & \lambda_n - t \end{vmatrix}$$

i.e. $$P_f(t) = \prod_{i=1}^{n} (\lambda_i - t) \,. \qquad [P.25]$$

But according to Proposition XXV (§. 5.8), λ_i is an eigen value iff λ_i is a solution of $P_f(t) = 0$. This establishes the theorem. ||

Problem 10

Show [235] of §.5.8.

Hint for the proof: Calculate the ij-th matrix entry of [235] by means of [234].

Problem 11

Prove that the set $\{y_1, \ldots, y_n\}$ of [299] is orthonormal

Proof

It is obvious, by inspection, that each y_i is normalized. Next, let us write [299] into more explicit forms:

$$y'_{n-1} = x_n - \sum_{i=1}^{n-1} \langle x_n \mid y_i \rangle y_i \qquad [P.26$$

and
$$y_n = y'_{n-1} \langle y'_{n-1}, y'_{n-1} \rangle^{-1/2}. \qquad [P.27$$

It is easy to see that $\langle y_1, y_2 \rangle = 0$ since

$$\langle y_1, y_2 \rangle = \langle x_1, x_1 \rangle^{-1/2} \langle y'_1, y'_1 \rangle^{-1/2} \langle x_1, y'_1 \rangle$$

and
$$\begin{aligned} \langle x_1, y_1 \rangle &= \langle x_1, x_2 \rangle - \langle x_2, y_1 \rangle^* \langle x_1, y_1 \rangle \\ &= \langle x_1, x_2 \rangle - \langle y_1, x_2 \rangle \langle x_1, y_2 \rangle \\ &= \langle x_1, x_2 \rangle - \langle x_1, x_1 \rangle^{-1} \langle x_1, x_2 \rangle \langle x_1, x_1 \rangle \\ &= 0 \ , \end{aligned}$$

where we used the defining expressions for y_1 and y_1'. Now, by the mathematical induction on n, assume that $\langle y_i, y_j\rangle = 0$ for $i < j < n$. We have, by [P.26] and [P.27],

$$\langle y_n, y_j\rangle = \langle y'_{n-1}, y'_{n-1}\rangle^{-1/2} \langle \left(x_n - \sum_{i=1}^{n-1} \langle x_n, y_i\rangle y_i\right), y_j\rangle .$$

But

$$\langle \left(x_n - \sum_{i=1}^{n-1} \langle x_n, y_i\rangle y_i\right), y_j\rangle$$

$$= \langle x_n, y_j\rangle - \sum_{i=1}^{n-1} \langle x_n, y_i\rangle \langle y_i, y_j\rangle$$

$$= \langle x_n, y_j\rangle - \sum_{i=1}^{n-1} \langle x_n, y_i\rangle \delta_{ij} = 0, \text{ if } j \neq n . \quad [P.28]$$

This completes the induction. ||

Problem 12

Show that [330] (§.5.11) makes V a G-module.

Proof

We have, for every $g, g' \varepsilon G$ and $x, x' \varepsilon V$,

i) $g \centerdot (x + x') = g^{\gamma}(x + x') = (g^{\gamma})x + (g^{\gamma})x' = g \centerdot x + g \centerdot x'$. [P.29]

ii) $(gg')\blacksquare x = (gg')^{\gamma} x = (g^{\gamma}(g')^{\gamma})x$

$= g^{\gamma}(g'\blacksquare x) = g\blacksquare(g'\blacksquare x)$. [P.30]

iii) $1_G \blacksquare x = (\hat{1}_V)x = x$. [P.31]

Hence V is a G-module under [330]. ||

CHAPTER VI

On Algebras

Summary

This chapter deals with elementary definitions and notions of algebras. We discuss both associative and non-associative algebras and the related concepts. The following topics are covered: ideals, quotient algebras, derivation mappings, free algebras, tensor algebras, exterior algebras, representations, extensions and Lie algebras.

In this chapter, except otherwise specified, R denotes a commutative ring with 1 and K denotes a field (a commutative division ring) of an arbitrary characteristic. M denotes an R-module. V denotes a K-module.

§. 6.1. Algebras, ideals and quotient algebras.

(def) An R-module $\mathcal{A}$ (with an R-module composition $\Omega_{\square}$) is called an R-algebra (or an algebra over R) with an algebra composition Ω_* if the mapping

$$\Omega_*: \mathcal{A} \times \mathcal{A} \longrightarrow \mathcal{A} \qquad [1]$$

with $\Omega_* : (x, x') \longrightarrow x * x'$, $x, x' \varepsilon \mathcal{A}$

satisfies the following conditions for every $r \varepsilon R$ and $x_i \varepsilon \mathcal{A}$:

i) $(x_1 + x_2) * x_3 = x_1 * x_3 + x_2 * x_3$ (left distributivity) [2

ii) $x_3 * (x_1 + x_2) = x_3 * x_1 + x_3 * x_2$ (right distributivity) [3

iii) $r \square (x_1 * x_2) = (r \square x_1) * x_2 = x_1 * (r \square x_2)$ (R-linearity) [4

Remark

We emphasize here that associativity w.r.t. Ω_* is not required for an algebra, i.e.

$$x_1 * (x_2 * x_3) \neq (x_1 * x_2) * x_3, \text{ in general.} \quad [5]$$

When it is necessary to emphasize this point some authors call an "algebra" (without modifiers) a "non-associative algebra" or "not-necessarily-associative algebra".

(def) A K-algebra $\mathcal{A}$, with an algebra composition $\Omega_\circ$, is calle an associative algebra if the composition $\Omega_\circ$ is associative, i. if

$$(x \circ x') \circ x'' = x \circ (x' \circ x''), \text{ for every } x,x',x'' \varepsilon \mathcal{A} \quad [6]$$

Remark

Axiom [6] implies: an R-module $\mathcal{A}$ is an associative algebr with an algebra composition $\Omega_\circ$ iff $\mathcal{A}$ is a ring w.r.t. $\Omega_\circ$.

Example

A handy but never-the-less important example is $\mathrm{End}_R V$. We know that $\mathrm{End}_R V$ is an R-module. If the algebra composition is defined as the composite mapping of any two R-endos then $\mathrm{End}_R V$ becomes an associative algebra.

(convention) AA = associative algebra. [7]

$$[\mathrm{End}_R V]_{AA} \equiv \mathrm{End}_R V \text{ as an associative algebra.}$$

Remark

An R-algebra $\mathcal{A}$ involves the following compositions:

i) R, as a commutative ring, is equipped with an addition and a multiplication; their units are denoted by 0 and 1, respectively.

ii) $\mathcal{A}$, as an R-module, has an addition and a module composition:

$$\Omega_+ : \mathcal{A} \times \mathcal{A} \longrightarrow \mathcal{A} \qquad [8]$$

and

$$\Omega_\square : R \times \mathcal{A} \longrightarrow \mathcal{A}. \qquad [9]$$

iii) $\mathcal{A}$, as an R-algebra, has an algebra composition,

$$\Omega_* : \mathcal{A} \times \mathcal{A} \longrightarrow \mathcal{A} . \qquad [10]$$

We note that we have used the same word "addition" for R as a commutative ring and for $\mathcal{A}$ as an R-module, for economy of language. The additions will be denoted by the same symbol + . The following elementary properties are obvious:

$$1 \,\square\, x = x, \quad 0 \,\square\, x = 0_{\mathcal{A}}, \quad r \,\square\, 0_{\mathcal{A}} = 0_{\mathcal{A}} \qquad [11]$$

and $$x * 0_{\mathcal{A}} = 0_{\mathcal{A}} = 0_{\mathcal{A}} * x , \qquad [12]$$

for every $x \in \mathcal{A}$ and $r \in R$. All the three relations in [11] were known before. We leave the proof of [12] to the reader (or see the Problem Section).

(def) an element $e \in \mathcal{A}$ is called a unit element of the algebr[a] if $e * x = x * e = x$, for every $x \in \mathcal{A}$.

It is important to note that not every algebra can have a unit element. We shall, in general, make no assumption of the existence of a unit in an algebra except explicitly stated.

(notation) $1_{\mathcal{A}} \equiv$ the unit of the algebra.

If $\mathcal{A}$ has a unit then the unit is unique; the proof is similar to that of the ring theory.

(def) A subset, B, of an R-algebra $\mathcal{A}$ (with algebra compositio[n]

Ω_*) is said to be a subalgebra of $\mathcal{A}$ if B itself is an R-algebra w.r.t. Ω_*. B is a proper subalgebra of $\mathcal{A}$ if $B \neq \mathcal{A}$.

(notation) we write $B \underset{\text{alg}}{\subset} \mathcal{A}$ if B is a subalgebra of $\mathcal{A}$.

Similarly, we write $B \underset{\text{alg}}{\subsetneq} \mathcal{A}$ if B is a proper subalgebra of $\mathcal{A}$.

(def) Let $\mathcal{A}$ be an algebra with Ω_*. A subalgebra B, of is called a left ideal of $\mathcal{A}$ if

$$\mathcal{A} * B \subset B \qquad [13]$$

i.e. $x * y \in B$, for every $x \in \mathcal{A}$ and $y \in B$. [14]

Similarly, a subalgebra B, of $\mathcal{A}$, is called a right ideal of $\mathcal{A}$ if

$$B * \mathcal{A} \subset B \qquad [15]$$

i.e. $y * x \in B$, for every $x \in \mathcal{A}$ and $y \in B$.

(def) A left ideal or a right ideal (of an algebra) is called a one-sided ideal (of an algebra).

(notation) $B \underset{\text{left}}{\subset} \mathcal{A}$ denotes that B is a left ideal of $\mathcal{A}$ (and a similar notation for a right ideal).

(def) A subalgebra B, of an algebra $\mathcal{A}$ (with Ω_*) is a two-sided ideal (i.e. "ideal" for short) if B is both a left ideal and a right ideal, of $\mathcal{A}$.

(notation) $B \vartriangleleft \mathcal{A}$ denotes that B is an (two-sided) ideal of $\mathcal{A}$. [16]

We now introduce the concept of quotient algebras. Let H be an ideal (i.e. two-sided ideal) of an R-algebra $\mathcal{A}$ (with Ω_*). Construct the set

$$\mathcal{A}/H = \{\bar{x} \mid \bar{x} = x \bmod H, \quad x \in \mathcal{A}\} \qquad [17]$$

where $\bar{x} \equiv x \bmod H$ is defined by

$$x \bmod H \equiv \{z \mid z = x + y, \ y \in H\} . \qquad [18]$$

Clearly, the set $\mathcal{A}/H$ is a quotient R-module with the addition and the module composition defined in the usual way:

$$(x \bmod H) + (x' \bmod H) = x + x' \bmod H \qquad [19]$$

and $\quad r(x \bmod H) = rx \bmod H$, [20]

for every $x, x' \in \mathcal{A}$ and $r \in R$. Then it is straightforword to verify that $\mathcal{A}/H$ is an R-algebra w.r.t. the algebra composition $\Omega_{\bar{*}}$ defined by, for every $x, x' \in \mathcal{A}$,

$$(x \bmod H) \,\bar{*}\, (x' \bmod H) = x * x' \bmod H . \qquad [21]$$

(def) Let H be an (two-sided) ideal of an R-algebra $\mathcal{A}$, then $\mathcal{A}/H$ is called the quotient algebra of $\mathcal{A}$ modulo H with its R-algebraic structure defined by [19], [20] and [21].

Remark

In order that the algebra composition $\Omega_{\bar{*}}$ of [21] is well-defined it is necessary to require H to be an (two-sided) ideal of $\mathcal{A}$. For instance, we have

$$(x \bmod H) \,\bar{*}\, (y \bmod H) = x * y \bmod H , \qquad [22]$$

for every $x \in \mathcal{A}$ and $y \in H$. But, on the other hand, it is obvious that

$$y \bmod H = 0 \bmod H \qquad [23]$$

since $y \in H$. Hence, by substituting [23] into the LHS of [22], we get

$$(x \bmod H) \,\bar{*}\, (0 \bmod H) = x * y \bmod H$$

i.e. $$x * 0 \bmod H = x * y \bmod H$$

or $$0 \bmod H = x * y \bmod H .$$

Therefore

$$x * y \in H . \qquad [24]$$

In other words H is a left ideal of $\mathcal{A}$. Similarly, one can show that H has to be a right ideal of $\mathcal{A}$. Consequently, H is an ideal (i.e. a two-sided ideal) of $\mathcal{A}$. For this reason, if H is only a subalgebra or a one-sided ideal of $\mathcal{A}$ (instead of a two-sided ideal) then $\mathcal{A}/H$ can have only the structure of a quotient R-module instead of a quotient R-algebra. ||

§. 6.2. Derivation mappings

In this section $\mathcal{A}$ denotes a K-algebra with an algebraic composition Ω_*. Further, "K-endos" or $\mathrm{End}_K\mathcal{A}$ denote K-endos w.r.t. the K-module structure of $\mathcal{A}$.

(def) A mapping $d \in \mathrm{End}_K\mathcal{A}$ is called a derivation mapping if it satisfies the condition

$$d(x * x') = (dx) * x' + x * (dx') , \qquad [25]$$

for every $x, x' \in \mathcal{A}$.

(notation) **Der.** $\mathcal{A}$ = the set of all derivation mappings on $\mathcal{A}$.

Since derivation mappings are K-endos we can define the sum of two derivation mappings d and d' by:

$$(d + d')\, x = dx + d'x \text{ , for every } x \in \mathcal{A} \text{ .} \qquad [26]$$

Proposition I

Der.$\mathcal{A}$ forms a K-module w.r.t. the addition given by [26], and the K-module composition defined by

$$(kd)x = k(dx) \text{ .}$$

Proof

For any $d_1, d_2 \in$ Der.$\mathcal{A}$ and $x_1, x_2 \in \mathcal{A}$, we have

$$(d_1 + d_2)(x_1 * x_2) = d_1(x_1 * x_2) + d_2(x_1 * x_2)$$

$$= (d_1x_1) * x_2 + x_1 * (d_1x_2) + (d_2x_1) * x_2 + x_1 * (d_2x_2)$$

$$= (d_1x_1 + d_2x_1) * x_2 + x_1 * (d_1x_2 + d_2x_2)$$

$$= ((d_1 + d_2)x_1) * x_2 + x_1 * ((d_1 + d_2)x_2) \text{ .}$$

i.e. $(d_1 + d_2) \in$ Der.$\mathcal{A}$. [27]

This shows the "closure" property of Der.$\mathcal{A}$, w.r.t. the addition defined by [26]. Other axioms of a K-module are also obviously satisfied by Der.$\mathcal{A}$. ||

Remark

We note that it is wrong to expect Der.$\mathcal{A}$ to be an associative algebra by simply defining the algebra composition to be the one that yields "composite mappings". In other words, let us now define the mapping

$$\Omega_\circ : \text{Der.}\mathcal{A} \times \text{Der.}\mathcal{A} \longrightarrow \text{Der.}\mathcal{A} \qquad [28]$$

with $\Omega_\circ : (d_1, d_2) \longmapsto d_1 \circ d_2$, for $d_1, d_2 \in \text{Der.}\mathcal{A}$

where $d_1 \circ d_2$ denotes, as usual, a composite mapping. Then, for every $x, x' \in \mathcal{A}$, we have

$$(d \circ d')(x * x') = d(d'(x * x')) = d((d'x) * x' + x * (d'x'))$$

$$= (d(d'x)) * x' + (d'x) * (dx') + (dx) * (d'x') + x * (d(d'x'))$$

$$= ((d \circ d')x) * x' + (d'x) * (dx') + (dx) * (d'x') + x * ((d \circ d')x')$$

which, in general, is not equal to

$$((d \circ d')x) * x' + x * ((d \circ d')x')$$

because the sum

$$(d'x) * (dx') + (dx) * (d'x') \neq 0, \text{ generally.} \qquad [29]$$

Therefore $\Omega_\circ$ is not a composition (i.e. derivations do not satisfy the closure property w.r.t. $\Omega_\circ$). However, we shall see that Der.$\mathcal{A}$ can be made into a Lie algebra (to be discussed in §.6.8).

<u>Proposition II</u> (the "Leibniz" rule)

For any $d \in \text{Der.}\mathcal{A}$ and $x, x' \in \mathcal{A}$, we have

$$d^n(x * x') = \sum_{j=o}^{n} {}_nC_j\,(d^j x) * (d^{n-j}x') \qquad [30]$$

where $d^o x \equiv x$ and $d^n = \underbrace{d \circ d \circ \cdots \circ d}_{n \text{ copies}}$. [31]

<u>Proof</u>

Use mathematical induction. [30] is obviously true for $n = 1$, since

$$d(x * x') = {}_1C_o\, x * (dx') + {}_1C_1 (dx) * x' = x * (dx') + (dx) * x' \ .$$

Next, we assume that [30] is true for $n = m$ by the induction hypothesis, i.e.

$$d^m(x * x') = \sum_{j=0}^{m} {}_mC_j\ (d^j x) * (d^{m-j}x') \qquad [32]$$

$$\text{i.e.} \qquad d^{m+1}(x * x') = \sum_{j=0}^{m} {}_mC_j\, d((d^j x) * (d^{m-j}x')) \qquad [33]$$

But,

$$d((d^j x) * (d^{m-j} x')) = (d(d^j x)) * (d^{m-j} x') + (d^j x) * d(d^{m-j} x')$$

$$= (d^{j+1} x) * (d^{m-j} x') + (d^j x) * (d^{m-j+1} x').$$

Hence [33] becomes

$$d^{m+1}(x * x') = \sum_{j=0}^{m} {}_mC_j (d^{j+1} x) * d^{m-j} x) + \sum_{j=0}^{m} {}_mC_j (d^j x) * (d^{m-j+1} x').$$

By changing the subscripts from j to $i-1$ for the first sum above, and j to i for the second sum above, we obtain

$$d^{m+1}(x * x') = \sum_{i=1}^{m+1} {}_mC_{i-1} (d^i x) * (d^{m-i+1} x') + \sum_{i=0}^{m} {}_mC_i (d^i x) * (d^{m-i+1} x')$$

$$= \sum_{i=1}^{m} ({}_mC_{i-1} + {}_mC_i)(d^i x) * (d^{m-i+1} x') +$$

$$+ {}_{m+1}C_{m+1}(d^{m+1} x) * (d^o x') + {}_mC_o (d^o x) * (d^{m+1} x').$$

But ${}_mC_o = {}_{m+1}C_o$ and ${}_mC_{i-1} + {}_mC_i = {}_{m+1}C_i$. Hence

$$d^{m+1}(x * x') = \sum_{i=1}^{m+1} {}_{m+1}C_i (d^i x) * (d^{m-i+1} x') + {}_{m+1}C_o (d^o x) * (d^{m+1} x')$$

$$= \sum_{i=0}^{m+1} {}_{m+1}C_i (d^i x) * (d^{m-i+1} x') . \qquad [34]$$

This completes the mathematical induction. ||

If we now assume that

$$\text{char.} K = 0 , \qquad [35]$$

then we can divide both the RHS and the LHS of [30] by n! :

$$\frac{1}{n!} d^n(x * x') = \sum_{j=0}^{n} \left(\frac{1}{j!} d^j x\right) * \left(\frac{1}{(n-j)!} d^{n-j} x'\right) . \qquad [36]$$

In other words, the LHS of [36] is well-defined by the RHS of [36]. Therefore, we can write down <u>formally</u> the series:

$$\sum_{n=0}^{\infty} \frac{d^n}{n!} = \hat{1} + d + \frac{1}{2!} d^2 + \frac{1}{3!} d^3 + \cdots , \qquad [37]$$

where $\hat{1}$ is the identity-endo on $\mathcal{A}$. We want to show that in the case of

$$\dim. \mathcal{A} < \infty \quad \text{and} \quad \underset{\sim}{K} = \mathbb{C} \text{ or } \mathbb{R} , \qquad [38]$$

the series [37] <u>converges</u> for <u>every</u> derivation mapping d .

Since dim.$\mathcal{A} \equiv m < \infty$, every $d \in$ Der.$\mathcal{A}$ is an $m \times m$ matrix with its entries in $\underset{\sim}{K}$ (= $\mathbb{C}$ or $\mathbb{R}$) in a base of $\mathcal{A}$. Denote by $\underset{\sim}{d}$ the matrix of d. Let M be an upper bound of the absolute value of the entries of $\underset{\sim}{d}$. [37] has the form

$$\sum_{n=0}^{\infty} \frac{\underset{\sim}{d}^n}{n!} = \underset{\sim}{1} + \frac{1}{2!} \underset{\sim}{d}^2 + \cdots . \qquad [39]$$

Clearly, d^n is again an $m \times m$ matrix. Let us denote its ij-th matrix entries by $(d_n)_{ij}$, $0 \leq n < \infty$. We want to show that

$$\left|(d_n)_{ij}\right| \leq (mM)^n . \qquad [40]$$

[40] is obviously true for $n = 0$. By the mathematical induction on n, let us assume that [40] holds for $n \equiv t$, i.e.

$$\left|(d_t)_{ij}\right| \leq (mM)^t .$$

Hence

$$\left|(d_{t+1})_{ij}\right| = \left| \sum_{k=1}^{m} (d_t)_{ik}(d)_{kj} \right|$$

$$\leq m(mM)^t M = (mM)^{t+1} \qquad [41]$$

which completes the induction. Consequently, by [39] and [40], we have

$$\sum_{n=0}^{\infty} \frac{\underset{\sim}{d}^n}{n!} \leq \sum_{n=0}^{\infty} \frac{(mM)^n}{n!} \quad . \qquad [42]$$

Since the RHS of [42] is known to be <u>convergent</u>, the convergence of [39] is established. ||

When the series [37] is convergent, e.g. under the conditions of [38], we shall write:

$$\sum_{n=0}^{\infty} \frac{d^n}{n!} = e^d \quad (\text{or } \exp.d) \ . \qquad [43]$$

§. 6.3. <u>Algebra homomorphisms</u>.

As "homomorphism" implies the preservation of some algebraic structure, it is natural to define an "algebra hom" as a mapping that preserves the K-algebra structures.

(def) Let $\mathcal{A}$ and $\mathcal{A}'$ be R-algebras (with compositions Ω_* and $\Omega_{*'}$), then a mapping $\alpha \in \text{Map}(\mathcal{A}, \mathcal{A}')$ is called an <u>algebra homomorphism</u> from $\mathcal{A}$ into $\mathcal{A}'$ if α is an <u>R-module hom</u> and if

$$\alpha(x * y) = (\alpha x) *' (\alpha y) \qquad [44]$$

for every $x \in \mathcal{A}$ and $y \in \mathcal{A}'$.

Remark

Similar to the case of group-homs, we shall use the notations like Hom(A, B), End.A , etc, to denote the set of all algebra-homs, the set of all algebra-endos, etc. We shall simply call algebra-homs "homs" (without the modifier "algebra") whenever there is no danger of ambiguity.

Because the important role played by the special case of associative algebras we introduce the following notations (cf. [7]):

(notation) AA-hom ≡ an associative-algebra hom.

AA-Hom ≡ the set of all AA-homs.

Proposition III

Let $\mathcal{A}$ be a K-algebra and $f \in End.\mathcal{A}$, then the exponential mapping e^f is an automorphism.

Proof

First we remind the reader that only finite-dimensional algebras are considered. Hence, in a base, f is an $n \times n$ matrix where $dim.\mathcal{A} = n < \infty$.

If $\lambda_1, \dots, \lambda_m$ are the set of all distinct eigen values of f then it follows from [227] of Proposition XXX, §.5.8, that $e^{\lambda_1}, \dots, e^{\lambda_m}$ are the set of all distinct eigen values of e^f with the same multiplicities. Hence, the exponential of any matrix f is <u>non-singular</u>. Therefore, f is an automorphism. ||

Proposition IV

Let $\mathcal{A}$ be a K-algebra and $d \in \text{Der}.\mathcal{A}$, then exp.d is an automorphism on $\mathcal{A}$.

Proof

For every x and $x' \in \mathcal{A}$, we have

$$e^d(x * x') = \sum_{m=0}^{n-1} \frac{1}{m!} d^m(x * x')$$

$$= \sum_{m=0}^{2(n-1)} \left[\left(\sum_{j=0}^{m} \frac{1}{j!} d^j x \right) * \left(\frac{1}{(m-j)!} d^{m-j} x' \right) \right]$$

$$= \left(\sum_{j=0}^{n-1} \frac{1}{j!} d^j x \right) * \left(\sum_{i=0}^{n-1} \frac{1}{i!} d^j x' \right)$$

$$= (e^d x) * (e^d x') . \qquad [45]$$

Therefore, it is clear that exp.d is an algebra-hom. By Proposition III, exp.d is an automorphism (i.e. algebra-auto). ||

§. 6.4. Free algebras

The notion of "free" objects, as we have seen in the previous chapters, shares some common features among different algebraic structures (monoids, groups etc). The definition of a "free' algebra follows the same characterization.

(def) Let S be a set, F an R-algebra with $\phi \in \text{Map}(S. F)$. Then the pair $\{F, \phi\}$ is called a free R-algebra on S if, for every given R-algebra B and $\psi \in \text{Map}(S, B)$,

$$\underset{1!}{\exists} f \in \text{Algebra-Hom}(F, B) : f \circ \phi = \psi \ , \qquad [46]$$

i.e. the following diagram is commutative:

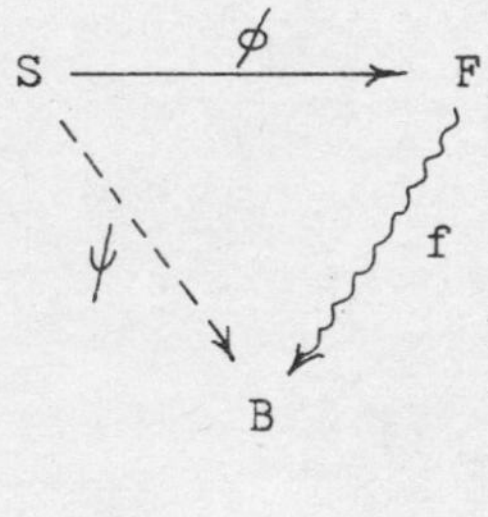

[47]

For economy of language, we shall often call F, instead of the pair $\{F, \phi\}$, a "free R-algebra" on S.

Similar to the case of a group (cf. §.2.8), we say that A' is a subalgebra generated by a subset A of an R-algebra $\mathcal{A}$ if A' is the intersection of all the subalgebras (of $\mathcal{A}$) containing A. Other definitions, like "generators", "the algebra generated by an arbitrary set" are similarly defined as for the groups, and we shall not repeat here.

Remark

If $\{F, \phi\}$ is the free algebra over a set S, then it is obvious (by definition) that $\phi(S)$ *generates* F or, we may also say that S **generates** F.

(def) A set S is said to *almost generate* an associative algebra B if $\phi(S) \cup 1_B$ generates B, where 1_B is the unit of B. Alternatively, we may also say that $\phi(s)$ *almost generates* B.

Analogous to the case of groups, we have the following isomorphism theorem.

Proposition V

Let $\{F_1, \phi_1\}$ and $\{F_2, \phi_2\}$ be two free R-algebras on the same set S, then

$$\underset{1!}{\exists} \eta \in \mathrm{Iso}(F_1, F_2) : \eta \circ \phi_1 = \phi_2 . \qquad [48]$$

Proof

The proof is similar to the case of free monoids or free groups (cf. §.1.4 and §.2.8). We leave the proof to the reader. ∎

Construction of a free algebra

By Proposition V, all the free algebras on the same set are isomorphic. Therefore, it is sufficient to give a particul construction.

Let S be the given set. To construct a free algebra on S, over the ground ring R, we proceed by the following steps.

Step 1

First let us introduce a sequence of disjoint sets startin from S. Define

$$S_1 \equiv S \ , \quad S_1' \equiv S_1 \times S_1 \qquad \text{(Cartesian product)}. \qquad [49]$$

Since S is an arbitrary given set it does not rule out the possibility that S may contain some cartesian products as its members. To avoid such a complication we introduce the new set S_2, S_3, ..., through a sequence of bijections in the following way: define S_2 such that

$$\exists h_1 \in \mathrm{Bij}(S_1', S_2) : S_2 \cap S_1 = \emptyset \ . \qquad [50]$$

Similarly, we introduce

$$S_2' \equiv (S_1 \times S_2) \cup (S_2 \times S_1) \qquad [51]$$

and define S_3 such that

$$\exists h_2 \in Bij(S_2, S_3) : S_3 \cap (S_1 \cup S_2) = \emptyset . \quad [52]$$

Proceed this inductively to define

$$S'_i \equiv \bigcup_{j+k=i+1} S_j \times S_k \quad [53]$$

and S_{i+1} such that

$$\exists h_i \in Bij(S'_i, S_{i+1}) : S_{i+1} \cap \left\{ \bigcup_{j=1}^{i} S_j \right\} = \emptyset . \quad [54]$$

In this way, the sequence of sets $\{S_i\}_{i=1,2,\ldots}$ thus constructed are mutually <u>disjoint</u>. Consider now the union

$$\bigcup_i S_i \equiv U . \quad [55]$$

It is obvious that $S \subset U$.

<u>Step 2.</u>

Construct the <u>free R-module</u> on U (cf. §.4.11) and denote this free module by $\{M, \mu\}$. Then $\mu(U)$ is a <u>base</u> of the free module M. M can now be made into an <u>R-algebra</u> by introducting the <u>algebra composition</u> Ω_*,

$$\Omega_* : M \times M \longrightarrow M \quad [56]$$

with $\Omega_*: (x, x') \longmapsto x * x' \equiv h_{i+j-1}(x, x') \in S_{i+j}$ [57]

for every $x \in S_i$, $x' \in S_j$. We leave to the reader the verification of the fact that $\{M, \phi\}$ is indeed a free R-algebra where ϕ is defined as the restriction,

$$\phi \equiv \mu\big|_S \ . \qquad [58]$$

We shall, in particular, look into the special case of the <u>free associative</u> R-algebra on a set S. From the preceding discussion, the definition of a <u>free associative algebra</u> needs no further elaboration. In this case, the algebra composition is required to be <u>associative</u> and the algebra-homs are actually AA-homs. The procedure of construction of a free associative algebra on S over R is given below:

<u>Step 1</u>.

Construct the <u>free monoid</u> generated by the set S, and denote this monoid by M. (We remind the reader that elements of M are called "monomials" or sometimes "words" formed by elements of S. A monomial is required to have a <u>finite</u> degree The "empty monomial" is denoted by 1_M).

<u>Step 2.</u>

Construct the associative R-algebra F on S in the following way:

a) Use M as a base to establish the R-module structure of F. Then each element of F is a formal R-sum of "monomials"; any element α of F is of the form

$$\alpha = \sum_i r_i x_i , \quad \text{with } r_i \in R \text{ and } x_i \in M \qquad [59]$$

where the addition is defined by

$$r_1 x + r_2 x = (r_1 + r_2)x , \qquad [60]$$

for every $x \in M$ and $r_i \in R$.

b) The AA-composition for F is just the composition induced by the "monomial " composition of M.

The following proposition justifies the above construction.

Proposition VI

Let F be as defined above and let ϕ be the inclusion mapping of S into F. Then the pair $\{F, \phi\}$ is a free associative R-algebra on S.

Proof

Step 1.

Let B be any associative R-algebra with 1. Denote its AA-composition by $\Omega_\circ$. Consider any mapping $g \in \mathrm{Map}(S, B)$.

Then g induces a mapping

$$\tilde{g} : M \longrightarrow B , \qquad [61]$$

with $\tilde{g} : \{s_1, \ldots, s_n\} \longmapsto g(s_1) \circ g(s_2) \circ \ldots \circ g(s_n)$, [62]

for every $s_i \varepsilon S$, and

$$\tilde{g} : 1_M \longmapsto 1_B , \qquad [63]$$

where 1_M and 1_B are the units of M and B. $\{s_1, \ldots, s_n\}$ denotes a "monomial" (which is an element of M).

Clearly,

$$\tilde{g}(x_1 * x_2) = \tilde{g}(x_1) \circ \tilde{g}(x_2), \quad \text{for every } x_i \varepsilon M . \qquad [64]$$

But M is a <u>base</u> for the <u>R-module structure</u> of F, thus $\tilde{g}$ induces an R-linear mapping, f,

$$f : F \longrightarrow B , \quad \text{with}$$

$$f : \sum_i r_i\{s_1, \ldots, s_{n_i}\} \longmapsto \sum_i r_i g(s_1) \circ g(s_2) \circ \cdots \circ g(s_{n_i}) .$$

f is clearly an AA-hom. ∎

<u>Step 2</u>.

Denote by B_1 the sub-AA of B <u>almost generated</u> by the set $\phi(S)$. Then, by [62] and [63], we have

$$f(M) \subset B_1$$

Hence $$f(F) \subset B_1 . \qquad [66]$$

On the other hand, it is obvious that

$$f(F) \supset (\phi(S) \cup 1_B) . \qquad [67]$$

[66] and [67] yield:

$$f(F) = B_1 \equiv \text{the sub-AA of B } \underline{\text{almost}} \text{ generated by } \phi(S). \qquad [68]$$

<u>Step 3</u>.

The uniqueness of f is rather obvious. For if there is an AA-hom,

$$f' : F \longrightarrow B,$$

with $$f' : 1_M \longmapsto 1_B$$

and $$f'\Big|_M = f\Big|_M ,$$

we have, by [68],

$$f'\Big|_F = f\Big|_F$$

i.e. $$f' = f \ . \qquad [69]$$

This establishes the uniqueness of f . ▮

§. 6.5. Tensor algebras and symmetric algebras

Given an R-module, an associative algebra called the "tensor algebra" can be constructed by means of tensor products. Similar to a free algebra, the construction of a tensor algebra is unique "up to an isomorphism".

(def) Let R be a commutative ring and M be an R-module, then a tensor algebra on M is defined as a pair $\{T, \phi\}$, where T is an associtive R-algebra with a unit and $\phi \in Map_R(M,T)$ such that, for any associative R-algebra B with a unit and any $\psi \in Hom_R(M, B)$,

$$\underset{1!}{\exists}\ f \in AA\text{-}Hom(T, B) : f \circ \phi = \psi \ , \qquad [70]$$

i.e. the following diagram is commutative:

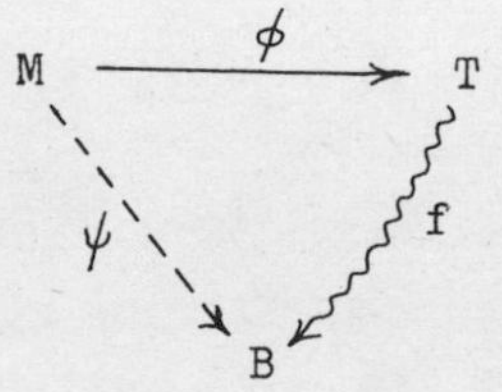

[71]

Proposition VII

Let $\{T_1, \phi_1\}$ and $\{T_2, \phi_2\}$ be two tensor algebras on an R-module M, Then

$$\exists_{1!} f \in \text{AA-Iso}(T_1, T_2) : f \circ \phi_1 = \phi_2 , \qquad [72]$$

i.e. the following diagram is commutative:

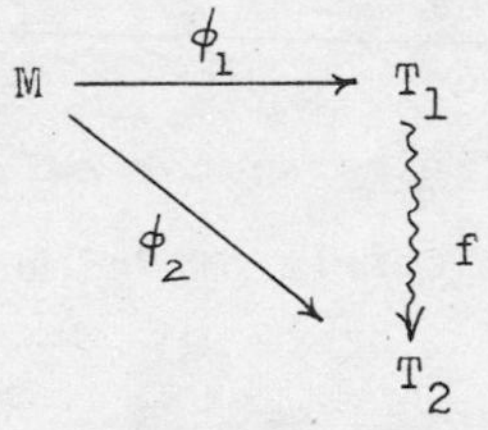

[73]

Proof

We leave this proof to the reader; it follows the same pattern as in the case of free monoids or free groups (cf. Chapters I and II). ||

Construction of a Tensor Algebra

Let M be an R-module. To construct a tensor algebra on M, we proceed as follows.

Step 1

Construct first the sequence of tensor products (of R-modules):

$$\{T_o, \tau_o\}, \{T_1, \tau_1\}, \{T_2, \tau_2\}, \ldots, \qquad [74]$$

where $T_i = \bigotimes^i M$ $(i = 0, 1, 2, \ldots)$ with $T_o = \bigotimes^o M = K$, $T_1 = \bigotimes^1 M = M$ and $\bigotimes^i M = M \otimes \cdots \otimes M$ (i copies, for $i \geq 2$). Then form the <u>direct sum</u> (of R-modules),

$$T = \bigoplus_{i=0}^{\infty} T_i, \qquad [75]$$

where, for each $\alpha \in T$, only <u>finite</u> number of components of α in T_i's are <u>non-zero</u> (in the spirit of "almost all zero").

<u>Step 2</u>

Define now the mappings $(i, j = 0, 1, 2, \ldots)$,

$$\tau_{i,j} : T_i \otimes T_j \longrightarrow T_{i+j} \quad \text{with} \qquad [76]$$

$$\tau_{i,j} : \{\tau_i(x_1, \ldots, x_i), \ {}_j(y_1, \ldots, y_j)\} \longmapsto$$

$$\tau_{i+j}(x_1, \ldots, x_i, y_1, \ldots, y_j), \qquad [77]$$

for every $x_k, y_k \in M$. Since

$$T \otimes T = \bigoplus_{i,j=0}^{\infty} T_i \otimes T_j, \qquad [78]$$

the mappings $\tau_{i,j}$ can be extended naturally to the domain $T \otimes T$. Denote the extended mapping (whose restriction to $T_i \otimes T_j$ are $\tau_{i,j}$) by τ. Define the mapping Ω_* by

$$\Omega_* : T \times T \longrightarrow T \quad \text{with} \tag{79}$$

$$\Omega_* : (t_1, t_2) \longmapsto \tau(t_1 \otimes t_2) , \tag{80}$$

for every $t_i \in T$. Then it is easy to see that T is an associative R-algebra with 1, w.r.t. Ω_*. The unit element of T is simply $\tau_0\{\emptyset\}$. We want to show that $\{T, \tau_1\}$ is a tensor algebra on M. τ_1, by definition, is obviously an R-mono. For any R-algebra B with 1 and any $\psi \in \mathrm{Hom}_R(M, B)$, we define the mapping

$$\psi_j : \underbrace{M \times \cdots \times M}_{j \text{ copies}} \longrightarrow B \quad \text{with} \tag{81}$$

$$\psi_j : (x_1, \ldots, x_j) \longmapsto (\psi x_1) \circ \cdots \circ (\psi x_j) , \tag{82}$$

for every $x_i \in M$. By the definition of a tensor product, for each j,

$$\underset{1!}{\exists} f_j \in \text{AA-Hom}(T_j, B) : f_j \circ \tau_j = \psi_j , \tag{83}$$

in virtue of the K-linearity of ψ_j. Denote by f the extension

of f_j's to the domain T in the natural way. Then it is straightforward to show that $f \in$ AA-Hom(T, B). Besides, the condition $f \circ \tau_1 = \psi$ is automatically satisfied as a consequen
Of [83] and definition [82]. The uniqueness of f follows easily from the fact that $\tau_1(M)$ almost generates T (i.e. $\tau_1($
together with the unit element generate T). **||**

(def) Let $\mathcal{S}$ be an associative R-algebra with composition $\Omega_\circ$ and $\phi \in \mathrm{Map}_R(M, \mathcal{S})$. Then the pair $\{\mathcal{S}, \phi\}$ is called a symmetric algebra on an R-module M if

i) $[\phi(s), \phi(s')] = \phi(s) \circ \phi(s') - \phi(s') \circ \phi(s) = 0$, [84]

for every $s, s' \in \mathcal{S}$.

ii) For any associative R-algebra B and any $\psi \in \mathrm{Hom}_R(M, B)$

$$\underset{1!}{\exists}\ f \in \mathrm{Hom}(\mathcal{S}, H) : f \circ \phi = \psi , \qquad [85]$$

i.e. the following diagram is commutative:

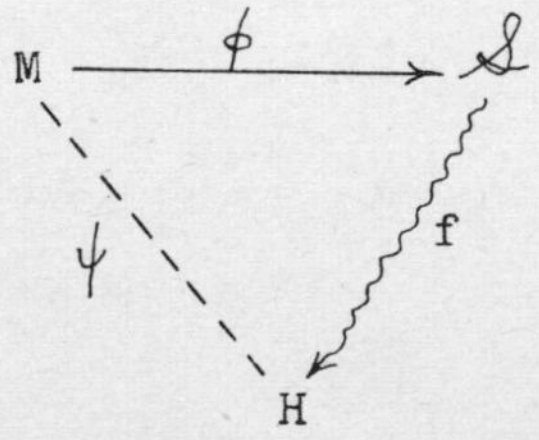

[86]

Construction of a Symmetric Algebra

Step 1.

Construct the tensor algebra $\{T, \alpha\}$.

Step 2.

Construct the two-sided ideal, J, of T, generated by the elements of the form (commutators):

$$[\eta(x), \eta(x')] \quad \text{with} \quad x, x' \in M .$$

Step 3.

Construct the quotient algebra T/J . Let β be the canonical mapping:

$$\beta : T \longrightarrow T/J ,$$

with $\beta : y \longmapsto y \text{ mod. } J, \quad \text{for every} \quad y \in T,$ [87]

and define $\phi \equiv \beta \circ \alpha$. Then it is straightforward to show that $\{T/J, \phi\}$ is a symmetric algebra.

§. 6.6. Exterior product and exterior algebra

M and N denote R-modules in this section.

To set the stage, we shall introduce some notation and

definitions. We shall denote by σ the permutation symbol on indices. For economy of notation, σ also denotes the following mapping:

$$\sigma : \overset{p}{\otimes} M \longrightarrow \overset{p}{\otimes} M \qquad (p = \text{a non-negative integer})$$

with $\sigma : v_1 \otimes \cdots \otimes v_p \longmapsto v_{\sigma(1)} \otimes \cdots \otimes v_{\sigma(p)}$ for $v_1 \in M$. [88]

(def) $f \in \mathrm{Hom}_R(\overset{p}{\otimes} M, N)$ is called a symmetric R-hom (or simply "f is symmetric") if

$$f \circ \sigma = f, \quad \text{for every permutation } \sigma, \qquad [89]$$

and is called an alternating R-hom (or "f is anti-symmetric") if

$$f \circ \sigma = (\mathrm{sgn}.\sigma) f, \quad \text{for every } \sigma, \qquad [90]$$

where $\mathrm{sgn}.\sigma$ denotes the "signature" of σ, i.e. $\mathrm{sgn}.\sigma = +1$ if the permutation is even and $\mathrm{sgn}.\sigma = -1$ if the permutation is odd.

(notation) $\text{s-Hom}_R(\otimes^p M, N)$ = the set of all symmetric R-homs. [91]

$\text{alt-Hom}_R(\otimes^p M, N)$ = the set of all alternating R-homs. [92]

For convenience, we define also the "symmetrizer" S_p,

$$S_p : x \longmapsto \sum_{\sigma} \sigma x \text{ , for every } x \in \otimes^p M \qquad [93]$$

and the "alternator" A_p ,

$$A_p : x \longmapsto \sum_{\sigma} (\text{sgn.}\sigma)\sigma x, \text{ for } x \in \otimes^p M . \qquad [94]$$

(notation) $\Lambda^p(M) \equiv (\otimes^p M)/\text{ker.}A_p$ [95]

$$\Lambda^o(M) \equiv R \text{ , } \Lambda^1(M) \equiv M .$$

(notation) $\overline{v_1 \otimes \cdots \otimes v_p} = v_1 \otimes \cdots \otimes v_p \text{ mod ker.}A_p$

<u>Proposition VIII</u>

$f \in \text{Hom}_R(\otimes^p M, N)$ is <u>alternating</u>

$$\Longleftrightarrow \text{ker.}A_p \subset \text{ker.}f \text{ .} \qquad [96]$$

<u>Proof</u>

<u>The necessity proof</u>

Since f is alternating, we have

$$f(z) = (\text{sgn.}\sigma)(f \circ \sigma)z \text{ , for every } z \in \text{ker.}A_p$$

i.e. $$\sum_{\sigma} f(z) = f[\sum_{\sigma}(\text{sgn.}\sigma)\sigma(z)]$$

$$= f A_p(z) = f(0) = 0 .$$

Hence $f(z) = 0$

i.e. $z \in \ker . f$.

Thus $\ker . A_p \subset \ker . f$. ∎

<u>The sufficiency proof</u>

By assumption, we have

$$\ker . A_p \subset \ker . f . \qquad [97]$$

For any σ and every $x \in \otimes^p M$,

$$A_p(x - (\mathrm{sgn}.\sigma)\sigma x)$$

$$= \sum_{\sigma'} (\mathrm{sgn}.\sigma')\sigma' x - \underbrace{\sum_{\sigma'} (\mathrm{sgn}.\sigma')(\mathrm{sgn}.\sigma)(\sigma' \circ \sigma)x}_{\sum_{\sigma''} (\mathrm{sgn}.\sigma'')\sigma''(x)} = 0 .$$

Therefore,

$$(x - (\mathrm{sgn}.\sigma)\sigma x) \in \ker . A_p \subset \ker . f \qquad (\text{by } [97])$$

i.e. $f(x - (\mathrm{sgn}.\sigma)\sigma x) = 0$

or $f(x) = (\mathrm{sgn}.\sigma)(f \circ \sigma)x$.

Hence f is alternating. ∎∎

Proposition IX

i) Define h to be the mapping

$$h : \otimes^p M \longrightarrow \Lambda^p(M) \ , \quad \text{with}$$

$$h : v_1 \otimes \cdots \otimes v_p \longmapsto \overline{v_1 \otimes \cdots \otimes v_p}, \text{ for every } v_i \in M. \qquad [98]$$

Then, for any given $f \in$ alt-$\mathrm{Hom}_R(\otimes^p M, N)$,

$$\underset{1!}{\exists} \; g \in \mathrm{Hom}_R(\Lambda^p(M), N) : f = g \circ h . \qquad [99]$$

ii) $$\text{alt-}\mathrm{Hom}_R(\otimes^p M, N) \longleftrightarrow \mathrm{Hom}_R(\Lambda^p(M), N) . \qquad [100]$$

Proof

i) By definition [98],

$$\ker.h \subset \ker.A_p \ , \qquad [101]$$

thus h is alternating, by Proposition VIII. The existence of g is rather trivial since we can take g to be

$$g : \overline{v_1 \otimes \cdots \otimes v_p} \longmapsto f(v_1 \otimes \cdots \otimes v_p) . \qquad [102]$$

Then $f = g \circ h$ follows immediately. The uniqueness of g is obvious. ∎

ii) By [99], we see that the mapping

$$\alpha : g \longmapsto g \circ h = f \qquad [103]$$

establishes an R-endo from $\mathrm{Hom}_R(\Lambda^p(M), N)$ to alt-$\mathrm{Hom}_R(\otimes^p M, N)$. Clearly, α is <u>surjective</u>. Further, since h is surjective, $f = 0$ iff $g = 0$. Hence α is an R-iso. ||

<u>Proposition X</u>

Defince h as in [98]. For every $f \in \mathrm{Hom}_R(N, M)$, denote by f^p the mapping

$$f^p : \otimes^p N \longrightarrow \otimes^p M \quad \text{with}$$

$$f^p : u_1 \otimes \cdots \otimes u_p \longmapsto (fu_1) \otimes \cdots \otimes (fu_p) , \qquad [104]$$

for every $u_i \in N$. Then $h \circ f^p$ is an <u>alternating</u> R-hom.

<u>Proof</u>

First, $h \circ f^p$ is obviously a R-hom. Since h is alternating (cf. [101]), we have

$$h \circ \sigma = (\mathrm{sgn.}\,\sigma) h$$

or

$$h \circ \sigma \circ f^p = (\mathrm{sgn.}\,\sigma) h \circ f^p$$

i.e.

$$h \circ f^p \circ \sigma = (\mathrm{sgn.}\,\sigma) h \circ f^p . \qquad [105]$$

Hence $h \circ f^p$ is alternating. ||

Proposition XI

Define h_* as the mapping

$$h_* : \otimes^p M \times \otimes^s M \longrightarrow \Lambda^{p+s}(M) \quad \text{with}$$

$$h_* : (v_1 \otimes \cdots \otimes v_p,\ v_{p+1} \otimes \cdots \otimes v_{p+s}) \longmapsto \overline{v_1 \otimes \cdots \otimes v_{p+s}}\,, \qquad [106]$$

for every $v_i \varepsilon M$. Then h_* is an alternating R-hom such that

$$h_*(x, y) = 0\,, \quad \text{for every } x \varepsilon \ker.A_p \text{ or } y \varepsilon \ker.A_s\,. \qquad [107]$$

Proof

For any $y \varepsilon \otimes^s M$, we define the following mapping

$$h_y : \otimes^p M \longrightarrow \Lambda^{p+s}(M) \quad \text{with}$$

$$h_y : x \longmapsto h_*(x, y)\,, \quad \text{for every } x \varepsilon \otimes^p M\,. \qquad [108]$$

One can show that h_y is an alternating R-hom (see the Problem Section). Hence, by Proposition VIII,

$$\ker.h_y \supset \ker.A_p\,, \qquad [109]$$

for every $y \varepsilon \otimes^s M$. On the other hand, by definition [109], we have

$$h_*(x, y) = h_y(x), \quad \text{for } x \varepsilon \otimes^p M \text{ and } y \varepsilon \otimes^s M\,. \qquad [110]$$

Thus, in virtue of [109],

$$h_*(x, y) = h_y(x) = 0 , \qquad [111]$$

for every $x \varepsilon \ker.A_p$ and $y \varepsilon \otimes^s M$.

Similarly we can define, for any fixed $x \varepsilon \otimes^p M$, the mapping

$$h^x : \otimes^s M \longrightarrow \Lambda^{p+s}(M) \quad \text{with}$$

$$h^x : y \longmapsto h_*(x, y), \quad \text{for every } y \varepsilon \otimes^s M . \qquad [112]$$

Again, h^x is an alternating R-hom. By a similar argument (cf.[111]), we conclude

$$h_*(x, y) = h^x(y) = 0, \quad \text{for every } y \varepsilon \ker.A_s,\ x \varepsilon \otimes^p M .$$ ||

(def) Exterior product

Define

$$\Omega_\wedge : \Lambda^p(M) \times \Lambda^s(M) \longrightarrow \Lambda^{p+s}(M) \quad \text{with}$$

$$\Omega_\wedge : (x \bmod \ker.A_p,\ y \bmod \ker.A_s) \longmapsto h_*(x, y) , \qquad [113]$$

for every $x \varepsilon \otimes^p M$ and $y \varepsilon \otimes^s M$, where h_* is the mapping defined by [106]; Proposition XI assures that the mapping $\Omega_\wedge$ is well-defined. For any

$\alpha \varepsilon \otimes^p(M)$ and $\beta \varepsilon \Lambda^s(M)$, we write

$$\Omega_\wedge : (\alpha, \beta) \longmapsto \alpha \wedge \beta , \qquad [114]$$

and call $\alpha\wedge\beta$ the exterior product (or Grassmann product)of α and β . In particular, we have, for any $v_i \varepsilon M$, the following different expressions for $v_1 \wedge \cdots \wedge v_p$:

$$v_1 \wedge \cdots \wedge v_p = \overline{v_1 \otimes \cdots \otimes v_p} = h(v_1 \otimes \cdots \otimes v_p)$$

$$= v_1 \otimes \cdots \otimes v_p \text{ mod ker.} A_p . \qquad [115]$$

It is also obvious that

$$(v_1 \wedge \cdots \wedge v_p) \wedge (v_1' \wedge \cdots \wedge v_s')$$

$$= v_1 \wedge \cdots \wedge v_p \wedge v_1' \wedge \cdots \wedge v_s' . \qquad [116]$$

Proposition XII

The exterior product is R-linear, associative and

$$x \wedge y = (-)^{ps} y \wedge x , \qquad [117]$$

for every $x \varepsilon \Lambda^p(M)$ and $y \varepsilon \Lambda^s(M)$.

Proof

The R-linearity and associativity of the exterior product are obvious; they amount to :

$$(r_1x_1 + r_2x_2) \wedge y = r_1(x_1 \wedge y) + r_2(x_2 \wedge y) \; , \qquad [118]$$

$$x \wedge (r_1y_1 + r_2y_2) = r_1(x \wedge y_1) + r_2(x \wedge y_2) \; , \qquad [119]$$

$$(x \wedge y) \wedge z = x \wedge (y \wedge z) \; , \qquad [120]$$

for every $r_i \in R$, $x_i \in \otimes^p M$, $y_i \in \otimes^s M$ and $z \in \otimes^t M$.

We shall only prove [117]. Since h is an alternating R-hom (see the statement below [101]), we have

$$h\sigma = (\text{sgn.}\sigma)\, h$$

or $\quad (\text{sgn.}\sigma)\, h\sigma(v_1 \otimes \cdots \otimes v_p) = h(v_1 \otimes \cdots \otimes v_p)$

i.e. $\quad (\text{sgn.}\sigma)\, v_{\sigma(1)} \wedge \cdots \wedge v_{\sigma(p)} = v_1 \wedge \cdots \wedge v_p \; . \qquad [121]$

Therefore, by R-linearity of the exterior product, [121] implies [117]. ‖

Proposition XIII

$$\Lambda^p(M) = 0 \;, \quad \text{if} \quad p > \text{dim.}M \; . \qquad [122]$$

Proof

Let $\{b_i\}_{i=1,\ldots,n}$ be a base of M ($\text{dim.}M \equiv n$). Then $\Lambda^p(M)$ has a base with elements

$$b_{i_1} \wedge \cdots \wedge b_{i_p} \ . \qquad [123]$$

This base is reduced to <u>zero</u> if $p > n$ since some factor must appear <u>more than once</u> in every exterior product [123]. Therefore, the whole base is reduced to zero if $p > n$. Consequently, [122] follows. ∎

(def) <u>Exterior algebra</u> (i.e. Grassmann algebra)

Let dim.M = n. Construct first the (K-module) direct-sum:

$$\Lambda(M) = \bigoplus_{p=0}^{n} \Lambda^p(M) \ . \qquad [124]$$

Then $\Lambda(M)$ can be made into an <u>associative</u> R-algebra by introducing the algebra composition $\Omega_\wedge$,

$$\Omega_\wedge : \Lambda(M) \times \Lambda(M) \longrightarrow \Lambda(M) \quad \text{with}$$

$$\Omega_\wedge : (x, y) \longmapsto x \wedge y \quad \text{such that} \qquad [125]$$

$$x \wedge y = \sum_{p,t=0}^{n} x^{(p)} \wedge y^{(t)} \ , \qquad [126]$$

where $x \equiv \sum_{p=0}^{n} x^{(p)}$, $y \equiv \sum_{p=0}^{n} y^{(p)}$ and $x^{(p)}, y^{(p)} \varepsilon \Lambda^{(p)}(M)$

The associative algebra $\Lambda(M)$, with composition $\Omega_\wedge$, is called the exterior algebra (i.e. Grassmann algebra) on M.

§. 6.7. Lie algebras

In this section $\mathcal{L}$ denotes a Lie algebra over a field K.

Lie algebra is discussed here as an important example of non-associative algebra. We shall use the compact notation $\star$ for a Lie algebra composition in place of the more conventional notation [,] which is reserved (as in the previous chapters) for the commutator between two elements.

(def) An algebra, $\mathcal{L}$ (with an algebra composition $\Omega_\star$), is called a Lie algebra if the mapping

$$\Omega_\star : \mathcal{L} \times \mathcal{L} \longrightarrow \mathcal{L}$$

with $\Omega_\star : (x, x') \longmapsto x \star x'$, for x and $x' \in \mathcal{L}$ [127]

satisfies the conditions:

i) $x \star x = 0$, (self-annihilation) [128]

ii) $\sum_{\text{cyclic}} x_1 \star (x_2 \star x_3) = 0$, (Jacobi identity) [129]

for every $x, x_i \in \mathcal{L}$. [129] has the meaning:

$$x_1 \star (x_2 \star x_3) + x_3 \star (x_1 \star x_2) + x_2 \star (x_3 \star x_1) = 0 \, . \qquad [130]$$

The composition $\Omega_\star$ will be referred to as a Lie-algebra composition.

Remarks

i) In contradistinction to an associative algebra, a unit element w.r.t. a Lie-algebra composition cannot be defined for any non-zero Lie algebra. Since if e is a unit element then $e \star e = 0$ implies that $e = 0$. But this is impossible if $\mathcal{L} \neq \{0\}$ because $x = x \star e = x \star 0 = 0$, for every $x \in \mathcal{L}$.

ii) We note that [128] implies "anti-symmetry"; i.e.

$$x_1 \star x_2 = - x_2 \star x_1 \text{ , for any } x_i \in \mathcal{L} . \qquad [131]$$

The proof is simple. By [128], we have

$$(x_1 + x_2) \star (x_1 + x_2) = 0$$

i.e.
$$x_1 \star x_1 + x_2 \star x_2 + x_1 \star x_2 + x_2 \star x_1 = 0 \qquad [132]$$

or
$$x_1 \star x_2 + x_2 \star x_1 = 0$$

where we used [128] for the first two terms of [132].

iii) If the ground field K of $\mathcal{L}$ has a characteristic not equal to 2, then [131] implies [128]. This is obvious since [131] yields, for every $x \in \mathcal{L}$,

$$x \star x = - x \star x$$

i.e. $$2(x \star x) = 0 . \qquad [133]$$

Since the characteristic of the ground field of $\mathcal{L}$ is not 2 (i.e. if $1+1 \neq 0$ where 0 and 1 denote the zero and the multiplicative unit of K), [133] implies that

$$x \star x = 0 . \qquad [134]$$

(abbreviation) LA = Lie algebra.

Examples

1) The set, $gl(n, \mathbb{C})$, of all $n \times n$ matrices with entries from $\mathbb{C}$ is a Lie algebra over $\mathbb{C}$. The K-module structure of $gl(n, \mathbb{C})$ is defined by the ordinary addition of matrices and multiplication of a matrix by a complex number. The Lie-algebra composition is defined by:

$$M_1 \star M_2 = M_1 \circ M_2 - M_2 \circ M_1 , \quad M_i \in gl(n, \mathbb{C}) \qquad [135]$$

where $M_1 \circ M_2$ is the ordinary matrix multiplication.

2) The set, $sl(n, \mathbb{C})$, of all $n \times n$ unimodular matrices (i.e. matrices with unit determinants) with complex entries is a Lie algebra over $\mathbb{C}$. Its K-module and Lie-algebra structures are defined in the same way as those of $gl(n, \mathbb{C})$.

3) Let B be an associative algebra (with $\Omega_\circ$ which is

associative, by definition). Define the mapping $\Omega_\star$ by

$$\Omega_\star : (x, x') \longmapsto x \star x' \equiv [x, x'] , \text{ for every } x, x' \in B, \qquad [136]$$

with $$[x, x'] \equiv x \circ x' - x' \circ x . \qquad [137]$$

Then B is a Lie-algebra w.r.t. the composition $\Omega_\star$. In other words, the associative algebra B now acquires a Lie-algebraic structure by the introduction of the "bracket operation" [,].

Here we have a very important example. Let V be a K-module, then we know that the set $End_K V$ forms an <u>associative</u> algebra. Therefore, $End_K V$ can be made into a <u>Lie algebra</u> by introducing the "bracket" composition [136].

(notations) $gl_K V \equiv$ the Lie-algebra structure of $End_K V$, under the "bracket" operation.

<u>Remark</u>

Since we can choose a <u>base</u> for any K-module V of dimension n, the following identification is trivial:

$$gl_K V = gl(n, K) . \qquad [138]$$

(def) A subset S, of a Lie algebra $\mathcal{L}$ (with composition $\Omega_\star$), is said to be a <u>stable subset</u> of $\mathcal{L}$ if

$$x_1 \star x_2 \subset \mathcal{L}', \quad \text{for every } x_i \in \mathcal{L}' \qquad [139]$$

i.e. $$\mathcal{L}' \star \mathcal{L}' \subset \mathcal{L}' \qquad [140]$$

where $$\mathcal{L}' \star \mathcal{L}' \equiv \{x \mid x = x_1 \star x_2, \; x_i \in \mathcal{L}\}. \qquad [141]$$

(def) A subset $\mathcal{L}'$, of a Lie algebra $\mathcal{L}$, is a sub Lie algebra (to be abbreviated by "sub-LA") if $\mathcal{L}'$ itself is a Lie algebra w.r.t. the induced Lie-algebra composition of $\mathcal{L}$.

(notation) we write $H \underset{LA}{\subset} \mathcal{L}$ if H is a <u>sub Lie algebra</u> of $\mathcal{L}$.

(def) A Lie algebra $\mathcal{L}$ is abelian (i.e. "commutative") if

$$\mathcal{L} \star \mathcal{L} = \{0\}.$$

<u>Example</u>

Let V be an R-mudule where R is a commutative ring. If we define $\Omega_\star$ by:

$$v_1 \star v_2 = 0, \quad \text{for every } v_1, v_2 \in V,$$

then this makes V an <u>abelian Lie algebra</u> w.r.t. $\Omega_\star$.

(def) B is an ideal of a Lie algebra $\mathcal{L}$ if B is a sub Lie algebra of $\mathcal{L}$ and if

$$B \star \mathcal{L} \subset B \qquad [142]$$

i.e. $b \star x \in B$, for every $b \in B$ and $x \in \mathcal{L}$. [143]

(notation) $B \lhd \mathcal{L}$ means that B is an ideal of $\mathcal{L}$.

We defined only (two-sided) "ideal" since "left ideal" and "right ideal" coincide in the case of a Lie algebra. By axioms of Lie algebra, we have

$$(x + x') \star (x + x') = 0 \ , \quad \text{for any} \quad x, x' \in \mathcal{L}$$

i.e. $x \star x + x' \star x' + x \star x' + x' \star x = 0$

or $x \star x' = - x' \star x$.

Hence $B \star \mathcal{L} = \mathcal{L} \star B$.

Since every ideal B (of an ideal) is two-sided, construct the "quotient Lie algebra modulo B", as in the general case of a non-associative algebra (cf. §.6.1).

(def) Let B be an ideal of $\mathcal{L}$, then the set $\mathcal{L}/B$, or sometimes written $\mathcal{L}$ mod B, defined by

$$\mathcal{L}/B = \{ \bar{x} \mid \bar{x} = x + B \equiv x \bmod B, \ x \in \mathcal{L} \} \qquad [144]$$

is called the quotient Lie algebra of $\mathcal{L}$ modulo B .

This definition is justified in §.6.1 for the general case of any algebra, and therefore will not be repeated here.

For a finite dimensional Lie algebra $\mathcal{L}$, we can choose a base (which is not unique) for $\mathcal{L}$:

$$X_1, X_2, \ldots, X_n .$$

Such a base can be characterized by the structure constants c_{ij}^k defined by:

$$X_i \star X_j = \sum_{k=1}^{n} c_{ij}^k X_k . \qquad [145]$$

Therefore, for different choices of bases for $\mathcal{L}$ we have different sets of structure constants. However, there are some common properties:

1) Anti-symmetry:

$$c_{ij}^k = - c_{ji}^k \qquad [146]$$

which follows directly from

$$X_i \star X_j = - X_j \star X_i . \blacksquare$$

2) Jacobi Identity:

$$\sum_{t=1}^{n}(c_{jk}^{t}\, c_{it}^{r} + c_{ki}^{t}\, c_{jt}^{r} + c_{ij}^{t}\, c_{kt}^{r}) = 0\ , \qquad [147]$$

where $i, j, k, r = 1, \ldots, n$.

<u>Proof</u>

We have

$$X_i \star (X_j \star X_k) = X_i \star \sum_t c_{jk}^{t} X_t$$

$$= \sum_{t,r} c_{jk}^{t}\, c_{it}^{r}\, X_r\ . \qquad [148]$$

Using the Jacobi identity [127],

$$\sum_{\text{cyclic}} X_i \star (X_j \star X_k) = 0\ , \qquad [149]$$

we obtain from [148]:

$$\sum_{\text{cyclic}} \sum_{t,r} c_{jk}^{t}\, c_{it}^{r}\, X_r = 0$$

i.e. $$\sum_r \Big(\sum_t \Big(\sum_{\text{cyclic}} c_{jk}^{t}\, c_{it}^{r} \Big)\Big) X_r = 0\ . \qquad [150]$$

Since X_r's form a <u>base</u>, they are K-linearly independent. Therefore

$$\sum_{t} \sum_{\text{cyclic}} c_{jk}^{t} \, c_{it}^{r} = 0$$

which establishes [147]. ∎

§. 6.8. Derivation mappings, solvability, nilpotency and semi-simplicity in Lie algebras.

In this section, $\mathcal{L}$ denotes a Lie algebra over K.

The definitions of Lie-algebra homs (to be abbreviated by LA-homs) and derivation mappings have already been covered, in §.6.2 and §.6.3, in the general discussion of non-associative algebras. There is no need of further elaboration of these definitions; we simply assume their specializations to Lie algebras in the present discussion.

The following property leads easily to the fact that Der.$\mathcal{L}$ is a Lie algebra w.r.t. the "bracket" operation.

Proposition XIV

$$[\text{Der.}\mathcal{L}, \text{Der.}\mathcal{L}] \subset \text{Der.}\mathcal{L} \,. \qquad [151]$$

Proof

For any $x_1, x_2 \in \mathcal{L}$, we have

$$[d_1, d_2](x_1 \star x_2) = (d_1 \circ d_2 - d_2 \circ d_1)(x_1 \star x_2)$$

$$= d_1((d_2x_1)\star x_2 + x_1\star(d_2x_2)) - d_2((d_1x_1)\star x_2 + x_1\star(d_1x_2))$$

$$= ((d_1\circ d_2)x_1)\star x_2 + x_1\star((d_1\circ d_2)x_2)$$
$$- ((d_2\circ d_1)x_1)\star x_2 - x_1\star((d_2\circ d_1)x_2)$$

$$= ([d_1, d_2]x_1)\star x_2 + x_1\star([d_1, d_2]x_2)$$

i.e. $[d_1, d_2] \in \mathrm{Der.}\mathcal{L}$. ||

Now it is easy to verify that Der.$\mathcal{L}$ is a <u>Lie algebra</u> w.r.t. the <u>bracket</u> operation. First, the preceding proposition establishes the "closure" property. Next it is trivial to see that, for every $k \in K$ and $d_i \in \mathrm{Der.}\mathcal{L}$,

i) $[(d_1 + d_2), d_3] = [d_1, d_3] + [d_2, d_3]$ [152]

ii) $[d_3, (d_1 + d_2)] = [d_3, d_1] + [d_3, d_2]$ [153]

iii) $k \square [d_1, d_2] = [(k \square d_1), d_2] = [d_1, (k \square d_2)]$ [154]

Furthermore, it is clear that

$$[d, d] = 0 \qquad [155]$$

and

$$\sum_{\text{cyclic}} [d_1, [d_2, d_3]] = 0 \ . \qquad [156]$$

Therefore Der.$\mathcal{L}$ is a <u>Lie algebra</u> w.r.t. the "bracket" composition.

It is trivial to see that

$$\mathrm{Der}.\mathcal{L} \underset{LA}{\subset} gl_K\mathcal{L}. \qquad [157]$$

We note that only the <u>K-module</u> structure of $\mathcal{L}$ is taken into account in $gl_K\mathcal{L}$.

We shall call Der.$\mathcal{L}$ the <u>Lie algebra of derivations</u> in $\mathcal{L}$ (it is sometimes referred to, by some authors, as the "derivation algebra" of $\mathcal{L}$).

<u>Remark</u>

For any <u>algebra</u> U (i.e. a non-associative algebra, in general) with an algebra composition Ω_*, one can show as exactly given in the preceding proposition that

$$[d_1, d_2](u_1 * u_2) = ([d_1, d_2]u_1) * u_2 + u_1 * ([d_1, d_2]u_2), \qquad [158]$$

therefore

$$d_1, d_2 \in \mathrm{Der.U} \Longrightarrow [d_1, d_2] \in \mathrm{Der.U}. \qquad [159]$$

Further, other properties given by the expressions [152] to [156] are also valid for any (non-associative) algebra U. Thus, similar to [157], we have the more general situation:

$$\mathrm{Der.U} \underset{LA}{\subset} gl_K V \ . \qquad [160]$$

Der.U is called "the Lie algebra of derivations in the (non-associative) algebra U".

An important special case of <u>derivation mappings</u> is the so-called "inner derivation mappings". We first introduce the very useful concept of "adjoint mappings":

(def) For any given $x \in \mathcal{L}$, the K-endo, denoted by adj.x, with

$$\mathrm{adj.x} : y \longmapsto x \star y, \text{ for every } y \in \mathcal{L}, \qquad [161]$$

is called an *adjoint mapping* in $\mathcal{L}$.

(def) adj.x, for $x \in \mathcal{L}$, is also called an *inner derivation*.

<u>Proposition XV</u>

adj.x is a derivation, for every $x \in \mathcal{L}$.

<u>Proof</u>

First, for every $x_i \in \mathcal{L}$, we have

$$(\mathrm{adj.x})(x_1 \star x_2) = x \star (x_1 \star x_2)$$

$$= - x_2 \star (x \star x_1) - x_1 \star (x_2 \star x)$$

$$= (x \star x_1) \star x_2 + x_1 \star (x \star x_2)$$

$$= ((\text{adj.}x)\,x_1) \star x_2 + x_1 \star ((\text{adj.}x)\,x_2) .$$

Next,

$$(\text{adj.}x)(x_1 + x_2) = x \star (x_1 + x_2)$$

$$= (x \star x_1) + (x \star x_2)$$

$$= (\text{adj.}x)\,x_1 + (\text{adj.}x)\,x_2 .$$

Hence adj.x ε Der.$\mathcal{L}$. ||

(notation) Inn.$\mathcal{L}$ = Adj.$\mathcal{L}$ = the set of all inner derivations on $\mathcal{L}$.

We leave the proof of the following property to the reader (or see the Problem Section):

<u>Proposition XVI</u>

$$\text{Inn.}\mathcal{L} \lhd \text{Der.}\mathcal{L} . \qquad [162]$$

i.e. Inn.$\mathcal{L}$ is an ideal of Der.$\mathcal{L}$, where Der.$\mathcal{L}$ is consider as the LA of derivations on $\mathcal{L}$ w.r.t. the bracket operation.

The above proposition allows us to construct the <u>quotient</u> Lie algebra of Der.$\mathcal{L}$ modulo Inn.$\mathcal{L}$:

$$\text{Der.}\mathcal{L}/\text{Inn.}\mathcal{L} \equiv \text{Out.}\mathcal{L}$$

which is often referred to as the outer derivation on $\mathcal{L}$.

Remark

It is straightforward to show that

$$\text{Out.}\mathcal{L} \subset \mathcal{L}. \tag{163}$$

(notations) For a Lie algebra $\mathcal{L}$ with composition $\Omega_\star$, we denote:

$$\mathcal{L}^{(0)} \equiv \mathcal{L}$$

$$\mathcal{L}^{(1)} \equiv \mathcal{L} \star \mathcal{L} \tag{164}$$

$$\mathcal{L}^{(2)} \equiv \mathcal{L}^{(1)} \star \mathcal{L}^{(1)}$$

and, in general,

$$\mathcal{L}^{(m)} \equiv \mathcal{L}^{(m-1)} \star \mathcal{L}^{(m-1)}, \quad m \in I_+ \tag{165}$$

where

$$\mathcal{L}^{(m)} \star \mathcal{L}^{(n)} \equiv \{ x \mid x = x' \star x'', x' \in \mathcal{L}^{(m)} \text{ and } x'' \in \mathcal{L}^{(n)} \}.$$

We also introduce the following notations:

$$\mathcal{L}^1 \equiv \mathcal{L}$$

$$\mathcal{L}^2 \equiv \mathcal{L}\star\mathcal{L} = \{x \mid x = x' \star x'', x', x'' \varepsilon \mathcal{L}\} \qquad [166]$$

$$\mathcal{L}^3 \equiv \mathcal{L}\star\mathcal{L}\star\mathcal{L} \equiv \mathcal{L}\star(\mathcal{L}\star\mathcal{L})$$

and, in general, (for simplicity of notation, we shall write $x \star y \star z$ in place of $x \star (y \star z)$ though it is not associative):

$$\mathcal{L}^m \equiv \underbrace{\mathcal{L}\star\cdots\star\mathcal{L}}_{m \text{ copies}}, \quad m \varepsilon I_+. \qquad [167]$$

Proposition XVII

$$\mathcal{L}^m \star \mathcal{L}^n \subset \mathcal{L}^{m+n}, \qquad [168]$$

for any positive integers m and n.

Proof

Use mathematical induction. Let the theorem be true for any fixed m and carry out the induction on the number n. First, [168] is obviously true for $n = 1$ which gives an identity. Next, let [168] be true for $n = t$. Hence

$$\mathcal{L}^m \star \mathcal{L}^t \subset \mathcal{L}^{m+t}$$

or $$\mathcal{L}\star\mathcal{L}^m \star \mathcal{L}^t \subset \mathcal{L}\star\mathcal{L}^{m+t}.$$

On the other hand, by Jacobi identity, we have

$$\mathcal{L}^m \star \mathcal{L}^t \star \mathcal{L} \subset \mathcal{L}^t \star \mathcal{L} \star \mathcal{L}^m + \mathcal{L} \star \mathcal{L}^m \star \mathcal{L}^t$$
$$\subset \mathcal{L}^{m+t+1} + \mathcal{L} \star \mathcal{L}^{m+t} \subset \mathcal{L}^{m+t+1}$$

which completes the induction. By the symmetry of m and n in [168] there is no need of carrying out an induction on m. This establishes the theorem. ∎

(def) A Lie algebra $\mathcal{L}$ is said to be solvable if

$$\exists\, m \in I_+ : \mathcal{L}^{(m)} = \{0\}. \qquad [169]$$

The minimal m satisfying [169] is called the solvability index of $\mathcal{L}$.

(def) A Lie algebra $\mathcal{L}$ is said to be nilpotent if

$$\exists\, n \in I_+ : \mathcal{L}^n = \{0\}. \qquad [170]$$

The minimal n satisfying [170] is called the nilpotency index of $\mathcal{L}$.

The following proposition shows that nilpotency is a stronger property than solvability:

Proposition XVIII

If a Lie algebra $\mathcal{L}$ is nilpotent then it is solvable, but the converse is not true.

Proof

i) For any $m \varepsilon I_+$,

$$\mathcal{L}^{(m)} = \{2^m \text{ copies of } \mathcal{L}, \text{ under LA-composition in the } \underline{\text{proper order}}\} \subset \mathcal{L}^{2m}, \quad [171]$$

in virtue of Proposition XVI (i.e. [168]). But the nilpotency of $\mathcal{L}$ implies that

$$\mathcal{L}^n = \{0\} \quad [172]$$

where $n =$ nilpotency index of $\mathcal{L}$. Therefore, for $2^m \geq n$, we have

$$\mathcal{L}^{(m)} \subset \mathcal{L}^{2m} = \{0\} \quad [173]$$

which establishes the solvability of $\mathcal{L}$. ∎

ii) We need only one counter-example to show that not every solvable Lie algebra is nilpotent. Consider now the two-dimensional Lie algebra $\mathcal{L}$ (over K) with a base $\{x, y\}$ such that

$$x \star y = cy \quad [174]$$

where c is a fixed **non-zero** element of K.

First, $\mathcal{L}$ is obviously solvable since

$$\mathcal{L}^{(1)} = (Kx + Ky) \star (Kx + Ky) = Ky$$

and $$\mathcal{L}^{(2)} = \mathcal{L}^{(1)} \star \mathcal{L}^{(1)} = Ky \star Ky = \{0\} . \qquad [175]$$

Next, we have

$$\mathcal{L}^2 = \mathcal{L}^{(1)} = Ky$$

and $$\mathcal{L}^3 = \mathcal{L} \star \mathcal{L}^2 = (Kx + Ky) \star Ky = Ky .$$

Therefore,

$$\mathcal{L}^2 = \mathcal{L}^3 = \mathcal{L}^4 = \cdots = Ky \qquad [176]$$

i.e. $$\nexists\, n \in I_+ : \mathcal{L}^n = \{0\} \qquad [177]$$

which establishes that $\mathcal{L}$ is not nilpotent. ||

Since the sum of solvable ideals is again solvable (we leave the proof to the reader) we can always find a solvable ideal B of a Lie algebra $\mathcal{L}$ such that B contains all the solvable ideals of $\mathcal{L}$. This leads to the definition:

(def) The radical of a Lie algebra $\mathcal{L}$ is the largest solvable of $\mathcal{L}$.

(def) A Lie algebra $\mathcal{L}$ is said to be simple if $\mathcal{L}$ is not abelian and if $\mathcal{L}$ does not contain any non-zero proper ideal.

(def) A Lie algebra $\mathcal{L}$ is said to be semi-simple if $\mathcal{L}$ does not contain any non-zero proper abelian ideal.

Remark

It is clear from the definitions that:

$$\text{simplicity} \; \underset{\not\Longleftarrow}{\Longrightarrow} \; \text{semi-simplicity}$$

For convenience of comparison, we summarize in the table below the definitions of simplicity and semi-simplicity for different algebraic structures:

	Simplicity	Semi-simplicity
group: (G)	does not contain any non-trivial, normal subgroup.	does not contain any non-trivial abelian normal subgroup.
ring: (R)	does not contain any non-trivial, two-sided ideal.	does not contain any non-trivial one-sided maximal ideal. (i.e. rad. $R = 0$)
module:	does not contain any non-trivial sub-module.	if every submodule can be supplemented.
AA: ($\mathcal{A}$)	does not contain any non-trivial, two-sided ideal.	does not contain any non-trivial one-sided maximal ideal. (i.e. rad.$\mathcal{A} = 0$)
LA: ($\mathcal{L}$)	does not contain any non-trivial ideal, and if $\mathcal{L}$ is not abelian.	does not contain any non-trivial solvable ideal (i.e. rad.$\mathcal{L} = 0$)

Note: By "non-trivial" we mean "non-zero and proper" for all the structures listed above except groups for which "non-trivial" means "$\neq \{1_G\}$ and proper".

§. 6.9. Extensions and representations.

The notions of exact sequences and extensions of algebras are parallel to those defined for monoids and groups. Many of the properties are common to the different structures, modulo some trivial modifications.

(def) Exact sequence

Let $\mathcal{A}_i$ be R-algebras with $\phi_i \in \text{Hom}(\mathcal{A}_i, \mathcal{A}_{i+1})$, $i = 1, \ldots, n-1$. Then the sequence

$$\mathcal{A}_1 \xrightarrow{\phi_1} \mathcal{A}_2 \xrightarrow{\phi_2} \cdots \xrightarrow{\phi_{n-1}} \mathcal{A}_n \qquad [178]$$

is called an exact sequence (of R-algebras and their homs) if

$$\text{im.}\,\phi_i = \text{ker.}\,\phi_{i+1}, \quad \text{for } i = 1, \ldots, n-2 . \qquad [179]$$

(def) An exact sequence of R-algebras and their hom,

$$B \xrightarrow{\eta} H \xrightarrow{\xi} \mathcal{A} \qquad [180]$$

is called an algebra extension of $\mathcal{A}$ by B if

$$\eta \in \text{Mon}(B, \mathcal{A}) \quad \text{and} \quad \xi \in \text{Epi}(H, \mathcal{A}) . \qquad [181]$$

Analogous to the situation in group theory, the statement [180] and [181] mean the same thing as the "exactness" of the following sequence:

$$0 \longrightarrow B \xrightarrow{\eta} H \xrightarrow{\xi} \mathcal{A} \longrightarrow 0 \qquad [182]$$

where the zeros (replaced the unit elements in group theory) are the trivial R-algebras $\{0\}$.

(def) two extensions,

$$B \xrightarrow{\eta} H \xrightarrow{\xi} \mathcal{A} \quad \text{and} \quad B \xrightarrow{\eta'} H' \xrightarrow{\xi'} \mathcal{A}, \qquad [183]$$

are said to be *equivalent* (or *isomorphic*) if there exists some $f \in \mathrm{Hom}(H, H')$ such that the following diagram is commutative:

$$\begin{array}{ccccc} & & H & & \\ & \overset{\eta}{\nearrow} & & \overset{\xi}{\searrow} & \\ B & & f \big\downarrow & & \mathcal{A} \\ & \underset{\eta'}{\searrow} & & \underset{\xi'}{\nearrow} & \\ & & H' & & \end{array} \qquad [184]$$

It is obvious to see that f is actually an *iso.*

For the general classification of extensions, we have the following definition, parallel to the situation in group theory.

(def) An algebra *extension*,

$$0 \longrightarrow B \xrightarrow{\eta} H \xrightarrow{\xi} \mathcal{A} \longrightarrow 0 \qquad [185]$$

is said to be *inessential* (or "split") if

$$\exists \mu \in \mathrm{Hom}(\mathcal{A}, H) : \xi \circ \mu = \hat{1}_{\mathcal{A}} \qquad [186]$$

where $\hat{1}_{\mathcal{A}}$ denotes the identity iso on $\mathcal{A}$.

(def) An algebra extension,

$$0 \longrightarrow B \xrightarrow{\eta} H \xrightarrow{\xi} \mathcal{A} \longrightarrow 0 \qquad [187]$$

is said to be trivial if

$$\exists \nu \in \mathrm{Epi}(H, B) : \nu \circ \eta = \hat{1}_H \qquad [188]$$

and $\exists \mu \in \mathrm{Mon}(\mathcal{A}, H) : \xi \circ \mu = \hat{1}_B$ [189]

where $\hat{1}_H$ and $\hat{1}_B$ are the identity-isos on H and B, respective

(def) An algebra extension

$$0 \longrightarrow B \xrightarrow{\eta} H \xrightarrow{\xi} \mathcal{A} \longrightarrow 0 \qquad [190]$$

is said to be a central extension if

$$\ker.\xi \subset \mathrm{cen}.H \qquad [191]$$

where cen.H, the center of the algebra H (with composition Ω_*) is defined by

$$\mathrm{cen}.H = \{ z \mid z \in H,\ z * h - h * z = 0 \text{ for every } h \in H \}. \qquad [192]$$

The specializations to different algebras are done by replacing:

"Hom" by "AA-Hom", for <u>associative</u> algebras,

"Hom" by "LA-Hom", for <u>Lie</u> algebras, etc.

In the remainder of this section, M denotes an R-module with module composition $\Omega_{\square}$.

As the representations of algebras are concerned, we shall only make some very brief remarks about associative algebras and Lie algebras.

(def) Let H be an associative algebra and $\gamma \in$ AA-Hom(H, $End_R M$). Then the pair $\{M, \gamma\}$ is called a *representation* of H (with an "underlying representation space" M). For economy of language, we shall simply refer to γ, instead of the pair $\{M, \gamma\}$, as a representation of H, if there is no danger of confusion.

(def) Let H be an associative algebra (with algebra composition $\Omega_\circ$). Then an R-module M is called a (left) *H-module* with an H-module composition $\Omega_\blacksquare$ if

$$\Omega_\blacksquare : H \times M \longrightarrow M \quad \text{with}$$

$$\Omega_\blacksquare : (a, x) \longmapsto a \blacksquare x, \quad \text{for every } a \in H \text{ and } x \in M, \qquad [193]$$

such that

$$\text{i)} \quad a \blacksquare (x + x') = a \blacksquare x + a \blacksquare x' \qquad [194]$$

ii) $(a + a') \blacksquare x = a \blacksquare x + a' \blacksquare x$ [195]

iii) $r \square (a \blacksquare x) = (r \square a) \blacksquare x = a \blacksquare (r \square x)$ [196]

iv) $(a \circ a') \blacksquare x = a \blacksquare (a' \blacksquare x)$, [197]

for every $a, a' \in H$; $x \in M$ and $r \in R$.

If we introduce the mapping

$$\gamma : H \longrightarrow \mathrm{End}_R M$$

with $\gamma : a \longmapsto a^\gamma$, for every $a \in H$, [198]

defined by

$$(a^\gamma)x \equiv a \blacksquare x \; , \quad [199]$$

Then the mapping γ is easily seen to be a representation of H with an underlying representation space M. The verification is rather straightforward. First, we have

$$a^\gamma (r \square x) = a \blacksquare (r \square x) = r \square (a \blacksquare x) = r \square (a^\gamma x) .$$

Next, we have

$$(a \circ a')^\gamma x = (a \circ a') \blacksquare x = a \blacksquare (a' \blacksquare x)$$

$$= a \blacksquare (a'^\gamma x) = (a^\gamma \circ a'^\gamma) x .$$

Hence $\gamma \in \text{AA-Hom}(H, \text{End}_R M)$.

In the case of a Lie algebra, representations are similarly defined, in view of [136] (or [138]) of §.6.7. A correspondence exists between a representation of a Lie algebra $\mathcal{L}$ and an $\mathcal{L}$-module. As before, we shall denote by $\mathcal{L}$ (with composition $\Omega_{\star}$) a Lie algebra over a field K.

(def) Let V be a (left) K-module, then every mapping

$$\gamma \in \text{Hom}(\mathcal{L}, gl_K V) \tag{199}$$

is called a representation of the Lie algebra $\mathcal{L}$ with an underlying representation space V. Here, Hom is the abbreviation for LA-Hom. In other words, the mapping

$$\gamma : \mathcal{L} \longrightarrow gl_K V \tag{200}$$

is a LA-hom. More explicitly, for every $x_1, x_2 \in \mathcal{L}$, we have

$$\gamma(x_1 \star x_2) = \gamma(x_1) \circ \gamma(x_2) - \gamma(x_2) \circ \gamma(x_1) \equiv [\gamma(x_1), \gamma(x_2)] . \tag{201}$$

For convenience we shall also use the notation:

$$\gamma(x) \equiv x^{\gamma} . \tag{202}$$

(def) A K-module V is called a (left) $\mathcal{L}$-module w.r.t. a module composition $\Omega_{\blacksquare}$ if

$$\Omega_{\blacksquare} : \mathcal{L} \times V \longrightarrow V \qquad \text{with}$$

$$\Omega_{\blacksquare} : (x, v) \longmapsto x \blacksquare v, \quad \text{for every} \quad x \in \mathcal{L}, \ v \in V, \qquad [203]$$

such that

$$\text{i)} \quad x \blacksquare (v + v') = x \blacksquare v + x \blacksquare v' \qquad [204]$$

$$\text{ii)} \quad (x + x') \blacksquare v = x \blacksquare v + x' \blacksquare v \qquad [205]$$

$$\text{iii)} \quad k \square (x \blacksquare v) = (k \square x) \blacksquare v = x \blacksquare (k \square v) \qquad [206]$$

$$\text{iv)} \quad (x_1 \star x_2) \blacksquare v = x_1 \blacksquare (x_2 \blacksquare v) - x_2 \blacksquare (x_1 \blacksquare v) \qquad [207]$$

for every $x, x' \in \mathcal{L}$, $k \in K$ and $v, v' \in V$. $\Omega_{\square}$ stands for the K-module composition of V.

Now , we shall construct the correspondence between a <u>representation</u> (of $\mathcal{L}$) and an <u>$\mathcal{L}$-module</u>. The required $\mathcal{L}$-module composition, denoted by $\Omega_{\blacksquare}$ can be defined as:

$$\Omega_{\blacksquare} : (x, v) \longmapsto x \blacksquare v, \qquad [208]$$

for every $x \in \mathcal{L}$ and $v \in V$, such that

$$x \blacksquare v = (x^{\gamma}) v. \qquad [209]$$

It is therefore clear, by [209], that to each representation there corresponds a uniquely determined $\mathcal{L}$-module. This works in both ways: an $\mathcal{L}$-module also uniquely fixes its corresponding representation. This is done in a trivial way. We simply define the corresponding representation γ to be such that, for any $x \in \mathcal{L}$,

$$x^\gamma : v \longmapsto (x^\gamma)v = x \blacksquare v, \quad \text{for each } v \in V . \qquad [210]$$

It is only routine to check that the γ defined by [210] is indeed a LA-hom; we have, for every $x, y \in \mathcal{L}$ and $v \in V$,

$$\begin{aligned}(x \star y)^\gamma : v \longmapsto (x \star y)^\gamma v &= (x \star y) \blacksquare v \\ &= x \blacksquare (y \blacksquare v) - y \blacksquare (x \blacksquare v) \\ &= (x^\gamma \circ y^\gamma - y^\gamma \circ x^\gamma)v\end{aligned}$$

i.e. $(x \star y)^\gamma = [x^\gamma, y^\gamma]$. ||

Problems with hints or solution for Chapter VI

Problem 1

Prove [12] of §. 6.1.

Proof

For any $k \in K$ and any $x, x' \in \mathcal{A}$, we have

$$k \,\square\, (x * x') = (k \,\square\, x) * x' \quad \text{(by axiom)}$$

and $$k \,\square\, (x' * x) = x' * (k \,\square\, x) \quad \text{(by axiom)}$$

In particular, when $k = 0$, we have

$$0 \,\square\, (x * x') = (0 \,\square\, x) * x' \qquad [P.1]$$

and $$0 \,\square\, (x' * x) = x' * (0 \,\square\, x) . \qquad [P.2]$$

Using the second relation of [11], we obtain from [P.1] and [P.2] respectively:

$$0_{\mathcal{A}} = 0_{\mathcal{A}} * x$$

and $$0_{\mathcal{A}} = x * 0_{\mathcal{A}} .$$ ||

Problem 2

Let d be a nilpotent derivation mapping on a K-algebra $\mathcal{A}$, then e^d is a K-iso on $\mathcal{A}$.

<u>Proof</u>

Let m be the nilpotency index of d, i.e.

$$d^i = \hat{0} \equiv \text{zero-endo on } \mathcal{A}, \quad \text{for any } i \geq m . \qquad [P.3]$$

It is obvious that, by [P.3],

$$e^d = \hat{1} + d + \frac{1}{2!} d^2 + \cdots + \frac{1}{(m-1)!} d^{m-1} \quad \text{(finite terms!)} \qquad [P.4]$$

$$\equiv \hat{1} + D \text{ , say.}$$

Nilpotency of d also leads to

$$D^m = \hat{0}$$

$$\text{i.e.} \quad (e^d)^{-1} = (\hat{1} + D)^{-1} = \hat{1} - D + D^2 + \cdots - \cdots (-)^{m-1} D^{m-1} . \qquad [P.5]$$

Since the RHS of [P.5] has only a <u>finite</u> number of terms, $(e^d)^{-1}$ converges. Hence $(e^d)^{-1}$ is well-defined. In other e^d is non-singular. Therefore, e^d is a K-iso. ▮

<u>Problem 3</u>

Show that the mapping h_y defined by [108] (Proposition XI, §. 6.6) is an alternating R-endo.

Proof

For any $v_i \varepsilon M$, we have

$$h_y\sigma : v_1 \otimes \cdots \otimes v_p \longmapsto h_y(v_{\sigma(1)} \otimes \cdots \otimes v_{\sigma(p)})$$

$$= h_*(v_{\sigma(1)} \otimes \cdots \otimes v_{\sigma(p)}, y) . \qquad [P.6]$$

Let y be written explicitly as $v_{p+1} \otimes \cdots \otimes v_{p+s}$, then

$$h_y\sigma(v_1 \otimes \cdots \otimes v_p) = \overline{v_{\sigma(1)} \otimes \cdots \otimes v_{\sigma(p)} \otimes v_{p+1} \otimes \cdots \otimes v_{p+s}} . \qquad [P.7]$$

On the other hand,

$$A_{p+s}(v_1 \otimes \cdots \otimes v_{p+s} - (\text{sgn.}\sigma)v_{\sigma(1)} \otimes \cdots \otimes v_{\sigma(p)} \otimes v_{p+1} \otimes \cdots \otimes v_{p+s})$$

$$= \sum_{\sigma'} (\text{sgn.}\sigma')v_{\sigma'(1)} \otimes \cdots \otimes v_{\sigma'(p+s)} -$$

$$- \sum_{\sigma'} (\text{sgn.}\sigma')(\text{sgn.}\sigma)v_{\sigma'\sigma(1)} \otimes \cdots \otimes v_{\sigma'\sigma(p)} \otimes v_{\sigma'(p+1)} \otimes \cdots$$

$$\cdots \otimes v_{\sigma'(p+s)} \qquad [P.8]$$

where σ' permutes <u>all</u> the $p+s$ indices while σ permutes only the <u>first</u> p indices above. We can extend the coverage of permutation of σ from p indices to $p+s$ indices in the follow trivial way:

$$\sigma(i) = i \,, \quad \text{for} \quad i > p \,. \qquad [P.9]$$

Then every term in the second sum of the RHS of [P.8] has uniform subscripts $\sigma'\sigma(i)$. Call $\sigma'' \equiv \sigma \circ \sigma'$. We have

$$\text{RHS of } [P.8] = \sum_{\sigma'} (\text{sgn.}\,\sigma') v_{\sigma'(1)} \otimes \cdots \otimes v_{\sigma'(p+s)} -$$

$$- \sum_{\sigma''} (\text{sgn.}\,\sigma'') v_{\sigma''(1)} \otimes \cdots \otimes v_{\sigma''(p+s)}$$

$$= 0 \,, \qquad [P.10]$$

where we made use of the fact that $\text{sgn.}\sigma'' = (\text{sgn.}\sigma)(\text{sgn.}\sigma')$. Hence

$$\text{LHS of } [P.8] = 0 \qquad [P.11]$$

i.e. $\quad A_{p+s}\{[h_y - (\text{sgn.}\sigma)h_y\sigma]v_1 \otimes \cdots \otimes v_p\} = 0$

or $\quad [h_y - (\text{sgn.}\sigma)h_y\sigma]v_1 \otimes \cdots \otimes v_p \;\varepsilon\; \text{ker.}A_{p+s} = \text{zero of } \Lambda^{p+s}(M)$

i.e. $\quad h_y = (\text{sgn.}\sigma)h_y\sigma$

or $\quad h_y\sigma = (\text{sgn.}\sigma)h_y \,. \qquad [P.12]$

Thus h_y is alternating. ∎

Problem 4

Prove Proposition XVI:

$$\text{Inn.}\mathcal{L} \subsetneq \text{Der.}\mathcal{L} . \tag{P.13}$$

Proof

First, we leave to the reader to prove that Inn.$\mathcal{L}$ is a sub-LA of Der.$\mathcal{L}$, i.e. to show:

$$[\text{Inn.}\mathcal{L} , \text{Inn.}\mathcal{L}] \subset \text{Inn.}\mathcal{L} . \tag{P.14}$$

Next, for any $\alpha \in \text{Inn.}\mathcal{L}$, we can write

$$\alpha = \text{adj.}x , \quad \text{for some} \quad x \in \mathcal{L} . \tag{P.15}$$

Consider any $\beta \in \text{Der.}\mathcal{L}$ and any $y \in \mathcal{L}$, we have

$$\begin{aligned}[\beta, \text{adj.}x]y &= (\beta \circ \text{adj.}x)y - (\text{adj.}x \circ \beta)y \\ &= \beta(x \star y) - x \star (\beta y) \\ &= (\beta x) \star y + x \star (\beta y) - x \star (\beta y) \\ &= (\text{adj}(\beta x))y\end{aligned}$$

i.e. $$[\beta, \text{adj.}x] = \text{adj}(\beta x) \in \text{Inn.}\mathcal{L} . \tag{P.16}$$

Thus,

$$[\text{Der.}\mathcal{L} , \text{Inn.}\mathcal{L}] \in \text{Inn.}\mathcal{L} . \ \| \tag{P.17}$$

Problem 5

Show that

$$\mathcal{L}^{(i)} \subsetneq \mathcal{L} \qquad [P.18]$$

and

$$\mathcal{L}^{(i)} \subsetneq \mathcal{L}^{(i-1)} \quad , \quad \text{for } i = 1, 2, \ldots \qquad [P.19]$$

Proof

We shall give only the proof of [P.18]. The proof of [P.19] is left to the reader.

Use mathematical induction on i. Assume that [P.18] is valid for i = n. Then

$$\mathcal{L} \star \mathcal{L}^{(n)} \subset \mathcal{L}^{(n)} \quad . \qquad [P.20]$$

On the other hand, by Jacobi identity, we have

$$\begin{aligned} \mathcal{L} \star \mathcal{L}^{(n+1)} &= \mathcal{L} \star \mathcal{L}^{(n)} \star \mathcal{L}^{(n)} \\ &\subset \mathcal{L}^{(n)} \star \mathcal{L} \star \mathcal{L}^{(n)} + \mathcal{L}^{(n)} \star \mathcal{L} \star \mathcal{L}^{(n)} \\ &= \mathcal{L}^{(n)} \star \mathcal{L} \star \mathcal{L}^{(n)} . \end{aligned} \qquad [P.21]$$

By applying [P.20] to the RHS of [P.21], we get

$$\mathcal{L} \star \mathcal{L}^{(n+1)} \subset \mathcal{L}^{(n)} \star \mathcal{L}^{(n)} = \mathcal{L}^{(n+1)} \qquad [P.22]$$

which completes the induction. ||

Problem 6

Let $\mathcal{L}$ be a solvable Lie algebra. Show that all its sub Lie algebras and homomorphic images are also solvable.

Proof

Let the solvability index of $\mathcal{L}$ be n, i.e.

$$\mathcal{L}^{(n)} = \{0\} . \qquad [P.23]$$

If B is any sub Lie algebra of $\mathcal{L}$, then it is clear that $\mathcal{L}^{(n)} \supset B^{(n)}$. Hence B is solvable. ▮

Next, if $\tilde{\mathcal{L}}$ is another Lie algebra and $f \in \text{Hom}(\mathcal{L}, \tilde{\mathcal{L}})$, then

$$f(\mathcal{L}^{(n)}) = (f\mathcal{L})^{(n)} \qquad [P.24]$$

and $$f(0) = \tilde{0} . \quad (\tilde{0} \equiv \text{zero of } \tilde{\mathcal{L}}) \qquad [P.25]$$

[P.24] and [P.25] yield

$$(f\mathcal{L})^{(n)} = \{\tilde{0}\} \qquad [P.26]$$

i.e. the homomorphic image of a solvable Lie algebra is also solvable. ▮▮

Problem 7

If a Lie algebra $\mathcal{L}$ contains a solvable ideal B such that the quotient algebra $\mathcal{L}/B$ is solvable, show that $\mathcal{L}$ itself is solvable.

Proof

Let n be the solvability index of $\mathcal{L}/B$, i.e.

$$(\mathcal{L}/B)^{(n)} = 0 \bmod B \ . \qquad [P.27]$$

Consider the "cononical isomorphism",

$$\eta \in \mathrm{Iso}(\mathcal{L}, \mathcal{L}/B)$$

with $\eta : x \longmapsto x \bmod B, \quad \text{for} \quad x \in \mathcal{L}$. [P.28]

We have

$$\eta(\mathcal{L}^{(n)}) = (\eta\mathcal{L})^{(n)} = (\mathcal{L}/B)^{(n)} = 0 \bmod B \ .$$

i.e. $\mathcal{L}^{(n)} \underset{\text{set}}{\subset} B$. [P.29]

On the other hand, B is solvable, by assumption. Let its solvability index be m, then

$$B^{(m)} = \{0\} \ . \qquad [P.30]$$

Thus, from [P.29],

$$\mathcal{L}^{(n+m)} = (\mathcal{L}^{(n)})^{(m)} \subset B^{(m)} = \{0\}$$

i.e. $\mathcal{L}$ is solvable. ||

Problem 8

Show that the intersection and the sum of two solvable ideals (in the same Lie algebra) are again solvable ideals.

Proof

i) If B_1 and B_2 are two solvable ideals in $\mathcal{L}$, then obviously

$$(B_1 \cap B_2) \lhd B_1 . \qquad [P.31$$

From Problem 6, the solvability of $B_1 \cap B_2$ follows.

ii) [P.31] allows us to construct the quotient algebra $B_1/(B_1 \cap B_2)$. On the other hand, we have obviously

$$B_2 \lhd (B_1 + B_2) \qquad [P.32$$

which allows us to construct the quotient algebra $(B_1 + B_2)/B_2$.

Since B_1 is solvable, its homomorphic image (under the usua canonical iso) $B_1/(B_1 \cap B_2)$ is also solvable. From the iso

$$\eta : (B_1 + B_2)/B_2 \longleftrightarrow B_1/(B_1 \cap B_2) \quad \text{with}$$

$$\eta : b_1 \bmod B_2 \longmapsto b_1 \bmod (B_1 \cap B_2), \quad \text{for } b_1 \in B_1 ,$$

we conclude that $(B_1 + B_2)/B_2$ is also solvable. Finally, the solvability of $B_1 + B_2$ follows from Problem 7 . ||

INDEX OF DEFINITIONS